THE NEW NATURALIST LIBRARY

A SURVEY OF BRITISH NATURAL HISTORY

MICROWORLDS

The aim of this series is to interest the general reader in the wildlife of Britain by recapturing the enquiring spirit of the old naturalists. The editors believe that the natural pride of the British public in the native flora and fauna, to which must be added concern for their conservation, is best fostered by maintaining a high standard of accuracy combined with clarity of exposition in presenting the results of modern scientific research.

THE NEW NATURALIST LIBRARY

MICROWORLDS

DAVID M. WILKINSON

WILLIAM
COLLINS

This edition published in 2026 by William Collins,
an imprint of HarperCollins*Publishers*

HarperCollins*Publishers*
1 London Bridge Street
London SE1 9GF

WilliamCollinsBooks.com

HarperCollins*Publishers*
Macken House, 39/40 Mayor Street Upper
Dublin 1, D01 C9W8, Ireland

First published 2026

A CIP catalogue record for this book is available
from the British Library.

Set in Nexus Serif Pro and Nexus Mix Pro

Edited and designed by
D & N Publishing
Baydon, Wiltshire

Printed in Bosnia-Herzegovina by GPS Group

Hardback
ISBN 978-0-00-871758-2

Paperback
ISBN 978-0-00-884109-6

Contents

Editors' Preface

In *Microworlds*, ecologist Dave Wilkinson, author of the first *New Naturalist* to take an explicitly ecological perspective in his *Ecology and Natural History*, breaks new ground once again. This is the first volume to put British natural history literally under the microscope. It recalls to mind a comment made to this writer by an Open University student who, looking up from his first view down a microscope at residential school, said 'Well, at my age, I never expected to see a whole new world.' That is exactly what you will find in this book – an unsuspected world with a rich natural history all of its own. And what a strange, marvellous world it is.

Single-celled creatures like ciliates, diatoms, amoebas and other protists live in aqueous habitats that may be as spacious as a puddle or as limited as the thin film of water trapped between the leaves of sphagnum moss. Under the microscope, they may pass with seemingly furious speed across the magnified field of view, but in truth they are battling with the viscosity of water that drags upon tiny creatures like a human swimmer in treacle. The microworld has all the physical laws that we are familiar with, but they operate at extremes, rendering rules of life that seem to have been re-written for the small scale. Gravity scarcely matters. Mouths, guts and anuses are superfluous. A whip-like flagellum is *de rigueur* and this mode of propulsion has evolved multiple times among protists. Without a horse, a whip will get you nowhere in our macroworld. (Of course we condemn animal cruelty.)

It was in the single-celled, microbial state of our long-ago ancestors that they acquired the bacterial symbionts that became mitochondria, without which multicellular life could not exist. The algal ancestors of plants acquired chloroplasts the same way. Most naturalists doubtless know this, but what you might not know is that in the whole history of life there appear to have been only two primary chloroplast-acquiring events. There is the one that occurred more than a billion and a half years ago that turned algae and their descendants green, and there was a second one just 120 million years ago. You can find that one in a unique amoeba called *Paulinella chromatophora*, quite likely to be living in a ditch near you.

The 'old naturalists', to whom the founding editors of *New Naturalist* paid homage in the manifesto that still guides us, were much more familiar with microscopes and the microworld than we are today. In Figure 44 on page 85, Professor Wilkinson resurrects the image of a late-nineteenth-century microscope slide that artfully displays a large assortment of diatoms, their glass shells of many shapes and sizes refracting a kaleidoscope of rainbow colours. That single image represents the ample promise of *Microworlds*. Even if you are fortunate enough to have 20:20 vision, this book should persuade you that every naturalist needs a microscope to fully appreciate the diversity of nature.

Author's Foreword and Acknowledgements

As large animals, we tend to have a highly biased and largely unrealistic view of life on Earth. Contrary to many people's idea of biodiversity, most of life's diversity is invisible to us as it is microscopic – so by definition requiring a microscope to see it. This means that its very existence is a relatively new discovery, as microscopes were not invented until the seventeenth century, and it was not until the nineteenth century that their technical quality really started to improve. The world of large organisms is also a geologically recent phenomenon, and for most of the history of life on Earth all life was microbial. A realistic TV documentary on 'Life on Earth' would be totally dominated by microbes.

My aim in writing this book is to open a door to the incredibly diverse microworlds that lie beyond our everyday perception. I will focus on what an earlier generation of biologists called algae and protozoa; these terms no longer feature in how we classify organisms, but they still make some ecological sense, as will be described later. This book therefore has relatively little to say about bacteria – except for cyanobacteria, which are ecologically algae – or about the wide range of microscopic animals, from nematodes to tardigrades. However, in various places I will compare algae and protozoa to bacteria, archaea and very small animals – as they share many similarities in their natural history. These algae and protozoa (often collectively called protists) also have the advantage of being typically larger than bacteria, and so easier to see with relatively simple microscopes. To further constrain this astonishing diversity to a more manageable size, I only consider the protists of soils and freshwater, and exclude all of the many parasitic species. I start my introduction to these organisms in the Prologue, by describing one of the microworlds of the Yorkshire Dales.

ACKNOWLEDGEMENTS

This book covers more biological diversity than most of the other books in the New Naturalist series put together. To attempt this, I have required considerable help. My editor, the plant ecologist and science writer Jonathan Silvertown, commented on the whole book in draft. Genoveva Esteban, Daniel Lahr and Hannah O'Regan commented on multiple chapters; while Tom Barker, Angela Creevy, Jane Fisher and Graeme Ruxton commented on drafts of individual chapters. They should all hopefully recognise their suggestions and improvements scattered throughout the book. Humphrey Smith and Richard Payne would undoubtedly have provided useful comments on several chapters had they lived longer – however, they both influenced my thinking on these topics, and in so doing their intellectual legacy lives on in this book.

Many people kindly helped with various points of detail. Namely: Hilary Atkinson, June Chatfield, Ford Doolittle, Jane Fisher, Liz Haworth, Alex Hyde, Andy Kesson, Dan Lahr, Enrique Lara, David Midgley, Edward Mitchell, Witold Morek, Anne-Marie Nuttall, Allison Perrigo, Tom Richards, Ferry Siemensma, Richard Small, Richard Thompson, Jenny Uglow and Jun Yang.

I am not an expert photomicroscopist, and so the fine photographs by Alex Hyde and Ferry Siemensma have greatly improved the book. Additional photos were taken by (or provided by) Hilary Atkinson, Steve Coulson, Angela Creevy, Genoveva Esteban, the late Bland Finlay, the late Wilhelm Foissner, Edward Mitchell, Janet Phillips, Tom Sherratt and Christine Strullu-Derrien (courtesy of the Trustees of the Natural History Museum London). All uncredited photographs are my own.

Turning a manuscript into a completed book is a large and skilled job. Myles Archibald at HarperCollins has been supportive and encouraging, as I wondered how to make an accessible, and readable, book on organisms that are unfamiliar to most people. I was very lucky in having Hugh Brazier as copy-editor. As with my previous New Naturalist, Hugh's input went well beyond sorting out typos and eccentric grammar, suggesting multiple improvements to the text. Sadly Hugh died as this book was going to press. His influence was widespread; many natural history books on my shelves contain authors' acknowledgements highlighting Hugh's editorial input; these form a lasting tribute to his abilities. Namrita and David Price-Goodfellow made a great job of the book design. Proofreading was by Jan McCann, and the index was compiled by Jan Ross. Robert Greenhalf's cover art beautifully illustrates that the microworld can be a source of artistic, as well as scientific, inspiration. At a time when much of publishing seems to suffer from cost cutting, the publisher's commitment to quality in the New Naturalist series, by employing so many talented professionals in the editing and design process, deserves particular praise.

PROLOGUE

Malham, Mites and Micrasterias

The sunlight of a still June evening slants through the wet woodland, picking out the detail of the bog-mosses growing on the peat. I am on the East Fen of Malham Tarn Moss in the Yorkshire Dales, taking photographs of moss-rich habitats to illustrate an article I am writing on testate amoebae (Figure 1). Some locations become well known not only because of the intrinsic interest of their natural history, but also because they have a long history of study. Such sites provide illustrations of how habitats have changed over the decades, and also tie you in to the history of the subject. Previous generations of scientists have studied such sites, providing examples that featured in lectures and reading back when I was a student. Tarn Moss – an expanse of raised bog and fen habitats just east of Malham Tarn – is one of these special places. I have known the Malham area since childhood, and I can remember reading the work of Margaret and Donald Pigott on the origin of the Moss when I was an undergraduate.[1] The field centre at Tarn House, only 10 minutes' walk from the edge of the Moss, provided convenient accommodation for generations of researchers and students studying the area from the late 1940s until its closure in 2022. It provided a base for people studying not only the Moss but also the microscopic plankton in Malham Tarn; some of their work will be described later in this book.

The reason I particularly wanted a photograph from Tarn Moss to illustrate an article on testate amoebae was because when I was a student in the early 1980s the most accessible identification key to these microbes for someone in Britain was a guide to common testate amoebae written by Sarah (known as Sally) Corbet.[2] This was based mainly on testates extracted from samples of bog-moss *Sphagnum* spp. from the Malham area. As Sally Corbet has since told

FIG 1. Carr woodland on Tarn Moss, Malham. The moss-rich ground cover makes a good habitat for testate amoebae and other microbes. Tarn Moss is the site that Sally Corbet used when collecting samples to create her key to testates living in *Sphagnum* in the early 1970s.

me, as a young academic she had to teach an introductory course on protozoa to undergraduates at Westerfield College, part of the University of London, and she was unhappy at the rather tedious laboratory sessions that she had inherited. When she saw testate amoebae from some Tarn Moss *Sphagnum* samples, while teaching on a course based at the field centre, she 'was bowled over by the beauty and interest of testates'. To use testates with beginning students she needed an accessible identification guide, so with advice from more experienced protozoologists (Jim Green and Bill Heal) she produced one and published it in the journal *Field Studies* in 1973.

Ten years later, as an undergraduate at Coventry Polytechnic, I too became fascinated by testates under the influence of Humphrey Smith, one of the ecology lecturers who had a particular research interest in Antarctic protozoa. My fascination continues, and explains why testate amoebae will feature extensively throughout this book. Following her one and only testate amoeba publication, Sally Corbet moved to the University of Cambridge and other research topics, becoming known for her research on insect pollination, as well as serving on the editorial board of the New Naturalist series for many years.

That June photo session was fitted into a few free hours while teaching on a first-year ecology field course for the University of Lincoln in 2022. Five years earlier I had also been there with first-year students and wanted an interesting topic for a half day free of formally assessed teaching. With a small group of students interested in microbes, I got permission to take water samples from some of the bog pools on Tarn Moss, along with small samples of the more aquatic *Sphagnum* mosses in the pools, to see what we could find. Back at the field centre we examined our catch using their microscopes, and it included several of the testate amoeba species described by Sally Corbet almost 45 years earlier. However, we had a range of other small organisms too. Not only did they help introduce the students to the microworld, but they can also usefully play the same role in this book.

The largest thing in our catch that day was a water mite only a few millimetres in length (Figure 2). That's large enough to see with the naked eye, but without at least a good hand lens it's impossible to make out much detail. Until the invention of the microscope anything much smaller than this was impossible to see, never mind study. Indeed, terrestrial mites were among the animals illustrated in the first book with extensive illustrations of organisms and other wonders drawn with the aid of microscopes, which was published in the seventeenth century.[3]

With only basic facilities at the field centre – relatively cheap microscopes with no inbuilt cameras – I photographed the mite by just pointing my compact camera down the microscope eyepiece. In the background, blurred as they are out of the plane of focus, are green algae. These are desmids, photosynthetic microbes which are particularly associated with acid waters such as those in bog

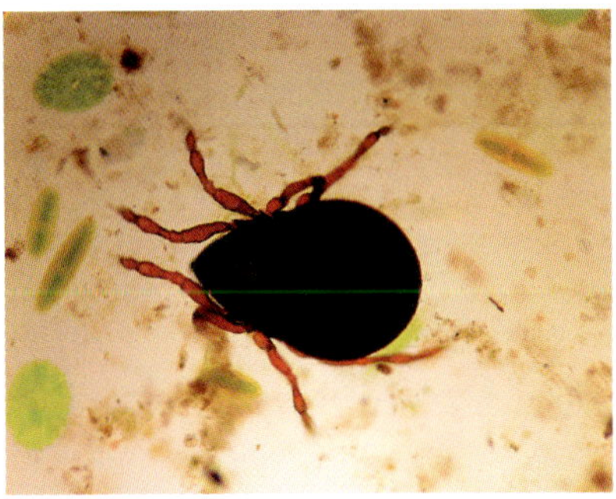

FIG 2. Water mite (only a couple of millimetres long) from a bog pool on Tarn Moss. The larger green shapes out of the plane of focus in the background are various types of desmid. The more circular alga in the upper left corner of the photograph is *Micrasterias*. This genus of desmid can have a diameter of up to 350 μm (0.35 mm). This means that an unaided eye could potentially see it as a green dot, but not resolve any detail – our eyes are capable of distinguishing things larger than about 0.1 mm (Dusenbery, 2009). Although not of the quality of some of the other photomicrographs in this book, it shows that you can take usable photos of tiny animals and true microbes with very simple equipment. In this case I used a compact camera to take the photograph through the eyepiece of a relatively basic compound microscope (a phone camera works just as well, as described in the Appendix).

surface pools. Noticeably smaller than the tiny mite, they were unknown until the development of microscopy, and yet they are large by the standards of many microorganisms. The relatively recent nature of our knowledge of these algae is nicely illustrated by the fact that in the early nineteenth century only four species of desmids had been described for Britain, but by the time John Ralfs published a classic monograph on British desmids in 1849 this had risen to 180 species – with Ralfs himself responsible for many of these new discoveries.[4] As Ralfs wrote in the opening sentence of his Introduction,

> *Until a recent period, the study of the minute objects which form the subject of this work have been more neglected in this kingdom than almost any other branch of Natural History.*[5]

In the top left corner of Figure 2 is the distinctive shape of a desmid in the genus *Micrasterias*. As Bellinger and Sigee noted in their guide to the identification of freshwater algae, desmids like *Micrasterias* have 'long been popular because of their beautiful shapes'[6] – which were nicely illustrated in Ralfs' monograph. As I hope the illustrations in this book show, the microworld is just as astonishingly beautiful as the more familiar wild flowers, birds and butterflies of the macroworld.

If we take the microworld to be defined by the organisms that appear in university-level textbooks of microbiology, then at the larger size end it overlaps with those accessible to the naked eye or hand lens – such as fungi and lichens. And indeed the buildings at Tarn House have an interesting collection of lichens growing on them, which made an accessible introduction to microbial ecology for students, allowing us to discuss the various algae and other microorganisms that are crucial to the lichens' function. At the other size extreme, many bacteria are so small as to be hard to see with the compound microscopes we used to look at the testates and desmids from the bog pools on Tarn Moss. Viruses are even smaller, and as parasitic 'genetic elements' they fall outside the remit of this book – indeed, it's hard to know if they should be considered alive at all.

Although this book ranges across these various microworlds, it concentrates on the middle size range, organisms we now tend to call protists – the algae and protozoa of an older generation of biologists. It also restricts itself to microbes living in terrestrial or freshwater habitats, and the free-living rather than parasitic species. Such organisms have crucial roles in many ecological processes. To introduce the basic structure of this microworld I start by describing selected scenes from the history of microbiology. My aim is not to give a systematic overview of the history of the subject, but to use the intellectual struggles of past scientists as an accessible way to introduce the basics of microbial natural history.

CHAPTER 1

The Hidden World of Infusoria

The date is 29 April 1669. After weeks of drought, it's just started raining again. Much of London is still a building site following the Great Fire of 1666, and has been a dusty place over the preceding period of dry weather.[1] At Arundel House, just off the Strand, a slightly built man is hurrying to a meeting – already somewhat stooped, despite being only in his mid-thirties. His posture imparts urgency to his movements, like the leaning purposeful figures moving through an L. S. Lowry painting. Since the Great Fire 'The Royal Society of London for Improving Natural Knowledge' has been using Arundel House as its meeting place, and the slight figure is Robert Hooke, the society's Curator of Experiments. He has to rush to make the meeting as he is hugely overcommitted – among many other projects, he is working alongside his friend Christopher Wren on the rebuilding of the city.

Today Hooke is showing the assembled Fellows an improved watch design, and also various biological samples to be viewed under a microscope. There has been some dispute with the Fellowship, who wanted him to bring a state-of-the-art microscope made by the instrument maker Christopher Cock of London to the meeting. But then, as now, complex scientific equipment often sulks and refuses to work when moved around too much, and Hooke didn't want to attempt to move Cock's 'Great Microscope' the half-mile from his rooms at Gresham College to Arundel House for the demonstration.

One of the samples Hooke shows, using a more easily transported microscope, is 'a kind of mould upon bookbinders' paste, which was found to have a fine moss growing on it, that had on top of its stems a head-like seed'. Hooke's 'fine moss' was actually a mould, a near-microscopic fungus. This world of microorganisms was completely new in the seventeenth century, and even today many people know little of microbes, except perhaps those which cause disease. In this chapter I use narratives from the history of microbiology to introduce the inhabitants of this often unfamiliar microworld.

HOOKE, VAN LEEUWENHOEK, AND THE START OF MICROBIOLOGY

A similar mould to that exhibited at Arundel House had been illustrated a few years before in Hooke's book *Micrographia*, published by the Royal Society in 1665, and demonstrated to various Fellows at a Society meeting in April 1663. Moulds, in Hooke's words 'microscopical mushrooms', were an accessible way into the microbial world. Microscopes had been around for some decades prior to this, and magnifying glasses much longer, but as far as we know neither had been used on what we would now call microorganisms.

A compound microscope can be thought of as a telescope used the other way round to make small things look large. It's easy to convince yourself this works: if you own a pair of binoculars (and most naturalists do), try looking at the letters in this sentence using binoculars the wrong way round so that the eyepiece becomes the objective lens, and you will see the text is magnified. Once Galileo had acquired working telescopes (around 1609) he quickly started making modified versions that worked better as basic microscopes, using them to look at the structures of bees and flies. Insects featured widely in these early studies; the earliest surviving published drawing made using a microscope is of bees from 1625, and the first botanical drawing, of very small seeds made with the aid of a microscope, comes from 1633.[2]

Telescopes quickly took off in scientific studies, likely because there was already a large body of theory on astronomy to which observations made using telescopes were highly relevant. Microscopy, however, was slower to develop.[3] After all, there was no existing microbial theory – which is hardly surprising, given that until the middle of the seventeenth century no one knew that microbes existed.

Things started to change a bit in London in 1665, and for a while microscopy became a fashionable obsession. This followed the publication of Hooke's *Micrographia*, the first well-illustrated book on microscopy – the subject's 'first all-round masterpiece' in the view of the scientifically inclined art historian Martin Kemp.[4] However, as described later in this chapter, it was the nineteenth century before microscopy really started to take off. Hooke's book also illustrated a range of invertebrates such as various flies, a louse, a pseudoscorpion, a mite and most famously a flea. The flea is spectacular – the drawing was too large and detailed to fit onto a single page of the book and so was printed as a fold-out illustration. Even today, opening out the paper to reveal the large flea is a memorable experience for anyone lucky enough to handle an original copy of *Micrographia*. At the time – when fleas were all too familiar but impossible to see in any detail with the naked eye – it must have been extraordinary. The spectacular sheet opens out to approximately 45 × 30 cm. On one occasion I was allowed to open the fold-out in an original copy. It was around 340 years old at the time and, although it was in surprisingly good condition, I unfolded it rather nervously.

It is these illustrations that made *Micrographia* famous, part state-of-the-art science and part coffee-table book. The engravings in the book were made from Hooke's own drawings. Such was his talent at drawing that his first job as a young teenager had been to be briefly apprenticed to the Dutch painter Peter Lely – recently arrived in London and not yet the famous artist he was later to become. In the case of the *Micrographia* flea it's possible that Hooke's friend Christopher Wren may also have been involved in making the original drawing.[5] Both Hooke and Wren used their considerable drawing skills in their joint architectural work, as well as in what we would now call science. The drawings of the flea and the other organisms don't appear to be illustrations of individual animals; they are thought to be mainly composite drawings based on looking at specimens from different angles, with varying types of illumination and often using more than one microscope.[6]

Arguably, however, the most important illustration was not the giant flea but the smaller illustration of a microscopic fungus – mould growing on the leather cover of one of Hooke's books (the upper illustration in Figure 3). It claims its place in history as the first illustration ever published of a microorganism, most probably a pin mould in the genus *Mucor*. As Hooke pointed out in his text, such moulds can be found on many 'kinds of putrif'd bodies', both animal and vegetable; a modern example is shown in Figure 4. The structures on stalks seen both in Hooke's drawing and in the modern photo show the spore dispersal structures, and in Hooke's drawing some of these have burst open to release spores into the air. However, this wasn't understood in the mid-seventeenth century, and Hooke wrote that he could find no sign of 'seeds' associated with these moulds, leading him to believe that they might 'be generated without seed' – that is, arise spontaneously from the rotting substrate.[7]

Micrographia gave Hooke a reputation across Europe. For example, a highly favourable review of the book was published in the important French *Journal des sçavans* (Journal of the learned). The review included small versions of two of Hooke's illustrations, which was highly unusual for *Journal des sçavans* at the time – one of these was Hooke's blue mould, while the other was a louse.[8]

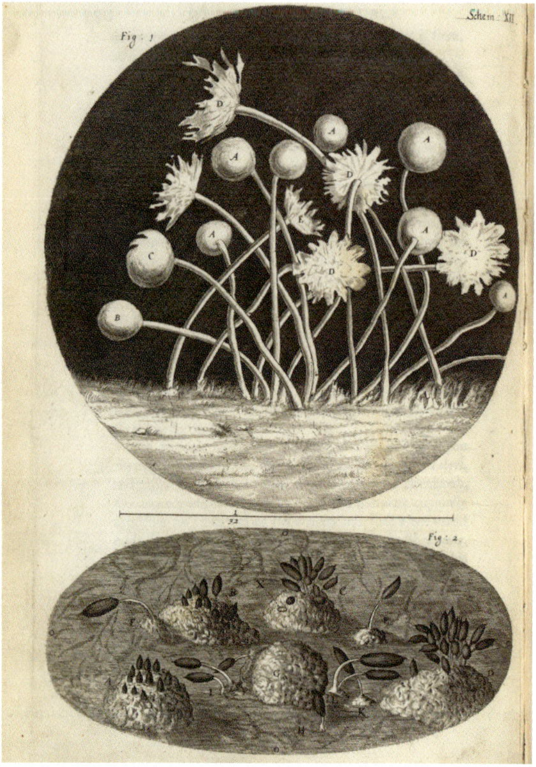

FIG 3. Illustrations of microfungi (described by Hooke as 'blue Mould') from *Micrographia* (Hooke, 1665) – the first published illustrations of microorganisms. The top illustration shows the mould Hooke found growing on leather book binding. We now know the reproductive structures (sporangia) are shown both before spore dispersal ('A' in Hooke's labelling) through to fully exploded sporangia that have released their spores (labelled 'D'). Hooke, understandably, didn't know what he was looking at and could see no sign of anything like 'seeds' that might spread the mould. The lower figure shows microbial sporangia on the surface of a rose leaf. Here, Hooke wondered if the structures were something like seed pods – they reminded him of what we now call capsules in mosses (from which the spores are spread). Hooke's scale bar (in the middle between the two engravings) is about 0.8 mm long. (Illustration from a copy of *Micrographia* in the Wellcome Collection, reproduced under a Creative Commons licence, CC BY 4.0)

The moulds illustrated by Robert Hooke are on the edge of what we now consider microorganisms, just visible to the naked eye but requiring magnification to see any detail of their structure. The first person to record completely microscopic organisms – totally invisible to the unaided eye – was Antoni van Leeuwenhoek of Delft, which was then part of the Dutch Republic. He was born in Delft in 1632 and lived there all his life except for being sent away for schooling followed by some years in Amsterdam in his late teens to learn the trade of a cloth merchant. He died in Delft at the advanced age of 90. It's likely he developed an interest in magnifying glasses and microscopes to look at the quality of woven cloth, but he started using homemade microscopes to look at a wide variety of other samples.[9]

FIG 4. Pin mould *Muccor* sp. (a microfungus) growing on the surface of a damaged tomato. (Alex Hyde)

The microscopes Hooke was using for demonstrations to the Royal Society were compound microscopes, most with two lenses; van Leeuwenhoek's microscopes made use of a single tiny lens made from small globules of glass, which are usually assumed to have been carefully polished, although attempts to recreate these microscopes show the polishing process may not be necessary. Van Leeuwenhoek was very secretive, so we are unsure exactly how he made his lenses, which were then mounted in a small metal plate with a viewing hole and screws to help focus the lens on the sample. The whole microscope was very small and handheld but could magnify a sample by at least 250 times. Hooke also knew how to make such microscopes, and described them in *Micrographia*, but he found them difficult to use, suggesting that although they gave clearer images for those 'whose eyes can well endure it' he found them 'offensive to my eye'.[10]

From 1673 onwards van Leeuwenhoek started sending letters describing his results to the Royal Society in London, many of which they published in their *Transactions*. His initial observations appear to have been attempts to replicate Hooke's work – including looking at mould. It seems very likely that he saw a copy of *Micrographia*, perhaps during a visit to London in 1668, and therefore got the idea for the construction of his single-lens microscopes from the description of these given in its preface.[11]

In September 1675 van Leeuwenhoek sent a letter to the Royal Society recording something really new. He described a lake about 10 km from Delft where he wondered why the water, which was clear in winter, was often coloured green in summer. He collected a water sample and took it home to look at with one of his microscopes. He found that in the water were 'several earthy particles, and some green streaks, spirally ranged'.[12] What he was describing for the first time, as 'green streaks', were filamentous algae, too small to be visible without the aid of a microscope. The mention of spirals suggests that he most likely was looking at the green algae *Spirogyra* – although it's all rather ambiguous. Many cyanobacteria are also filamentous, and often form blooms that colour the water during the summer, so these are also a possibility (although the green spirals sound very *Spirogyra*-like). He also mentions seeing other 'animalcula', what later biologists were to call protozoa. Some of these had 'green in the middle'. It has been suggested that these were the flagellate protist *Euglena viridis* – John Corliss was happy with this identification, calling it a 'clear, if brief, description',[13] but as with *Spirogyra* I think the exact identification has to be a bit of a guess. Alas, van Leeuwenhoek was no artist, so we have no illustrations of what he saw on this occasion, although in some later cases he hired local artists to try and draw what he was seeing.

Probably the most famous of van Leeuwenhoek's letters to the Royal Society is his 1677 contribution 'Concerning little animals', often referred to as his 'letter on protozoa'.[14] He had looked at samples of the spice pepper left to infuse in snowmelt water (the purest water easily available to him) and found to his 'great wonder, an incredible number of very little animals'. He described several creatures that appear to be protozoa (i.e. heterotrophic protists) and some others that were 'incredibly small' – he estimated that 100 of them laid end-to-end would not be as long as a grain of sand (he had no good way of measuring things viewed with his microscopes). There is a general consensus that these 'incredibly small' organisms were almost certainly bacteria, and even if they were not, then the tiny organisms he found in scrapings from his own teeth in 1683 were certainly bacterial.[15]

Van Leeuwenhoek's findings were not just taken on trust by the Royal Society. With considerable difficulty, Hooke managed to replicate the results in London and show these 'very little animals' to Fellows of the Society. A remarkable aspect of this interaction between van Leeuwenhoek and the Royal Society is that during its course there were several Anglo-Dutch Wars – indeed, for a time in 1667 the Secretary of the Royal Society was imprisoned in the Tower of London as a potential spy because of his extensive correspondence with scholars across Europe, irrespective of any ongoing hostilities.[16] Over the following centuries there has been considerable scepticism that you could really see things as small as bacteria with these early microscopes. However, studies by Brian J. Ford using surviving 'Leeuwenhoek' microscopes, along with replicas of the early microscopes, have shown that you can indeed see such small organisms using them.[17]

'THE ANIMALCULES OF INFUSIONS'

It might seem odd that some of van Leeuwenhoek's most famous microbiological observations were made on infusions of pepper in water. However, there was a good reason for using pepper that has nothing to do with looking for tiny organisms. He was interested in trying to work out why this spice tasted so distinctive, and wondered if it was anything to do with the pepper's shape; he thought that soaking it in water might make the pepper soft and easier to observe with his microscope. After finding strange microscopic creatures in the water infused with pepper, infusions of a variety of plant material became a standard way of culturing microorganisms for viewing under microscopes.[18] Hay was often used in such infusions, being readily available in a world where horse transport was so important, and this approach led to such organisms becoming known as infusoria.

Although now obsolete in biology, a web search quickly reveals that the term *infusoria* still survives, used by some aquarists for tiny organisms reared for fish food by soaking vegetation in water and waiting some days until good populations appear. Although I knew about the hay-infusion approaches from books on the history of microbiology, I had never tried to replicate these early experiments myself until starting to write this chapter. To see what happens, I took some dry grass from part of my garden managed as meadow vegetation and soaked it in rainwater. I deliberately chose to look at the samples using a rather basic microscope (an old compound microscope that had been marketed for amateur use over 50 years ago, which I had inherited from my father). I wanted to try and get more of a feel for what earlier microscopists would have seen, rather than using a modern state-of-the-art lab microscope.

After a couple of days, I tried sucking up a small amount of liquid from the bottom of the sample tube containing the infusion, placed it on a microscope slide and looked at it under magnifications of ×100 and ×400. My microscope showed a number of small organisms, apparently moving very quickly. Their movement made it clear that I was looking at living things, but also made it very hard to see what exactly I was seeing. Today you can freeze the movement with photography to get a clear look, but that wasn't an option for the early microscopists. Even when photomicrography started to develop in the nineteenth century the necessary exposure times made it largely useless for moving organisms. However, the movement that makes it hard to see the detail of what you are looking at is also what caused early microscopists to refer to the organisms as animalcules – little animals – as animals move around while plants stay put. With the benefit of modern knowledge it was obvious to me that the

OPPOSITE: **FIG 5.** Infusoria from Griffith & Henfrey (1875). Most of the organisms illustrated are ciliates, including a mix of species which are attached to the substrate by a stalk, and those that can move freely – either through the water or by gliding across the substrate.

'animalcules' I was looking at were mainly ciliate protists (Figure 5). But without that knowledge – that is, in the same position as Hooke or van Leeuwenhoek – I would have found it very difficult to work out what I was seeing. The number of microbes in my infusion increased over the next few days. Hooke found this too in his initial attempts to replicate van Leeuwenhoek's results. At first he failed to find any animalcules, but when he let his infusions stand for five or six days and looked again he wrote that 'then much to wonder I discovered vast multitudes of those exceeding small creatures'.[19]

In the seventeenth century it was really difficult to be sure what you were seeing when looking at infusoria. Although Hooke wrote that what you needed to see and draw the objects under your microscope was 'a sincere Hand, and a faithful Eye', in practice he found it much harder. The images were ambiguous, and interpretation varied from observer to observer.[20] There was a real worry that what you were seeing was an artefact of the technology, rather than showing you anything reliable about nature. For example, the seventeenth-century philosopher Margaret Cavendish strongly argued that you could only trust human senses unenhanced by technology, so microscopes could not be trusted.[21] It is certainly the case that some of what you see down a microscope can be rather misleading. Those fast-moving ciliates I saw in the hay infusion are actually moving very slowly. They look fast because you are fooled by the high magnification of the microscope; the distances they are travelling as you watch them apparently zooming around are in fact tiny. A typical fast speed for a ciliate protozoan is around 1 mm per second; that's 6 cm over the course of a minute.[22] As the slime mould expert John Tyler Bonner reminisced in old age, reflecting on his early views of apparently fast-moving protozoa: 'It was rather disappointing to hear that the microworld was not dazzlingly fast but rather sedate.'[23]

PLANTS, ANIMALS, OR SOMETHING ELSE?

Once the unsuspected world of infusoria had been discovered, you might think it would rapidly become an area of sustained scientific interest. But this didn't happen, and it had to wait for technical improvements in microscope design in the early nineteenth century before microscopy really began to become established as a research method in biology.[24] However, once large numbers of biological

INFUSORIA.
Pl. 25.
Tuffen West del.
London John Van Voorst 1873

studies using microscopes started to accumulate, the longstanding problems about the definition of plants and animals became ever harder to ignore.

As so often in the Western intellectual tradition, the problem can be traced back to Aristotle, over two millennia ago. In the view of the philosopher Mary Midgley, who was 'steeped in Aristotle' herself, 'he stands out as *the* biologist among philosophers'.[25] Aristotle had a strong interest in what we would now call marine biology, and several of the organisms he found on the coast – or dredged up by fishermen – puzzled him. In his *History of Animals* he wrote that 'in the sea, there are certain objects concerning which one would be at a loss to determine whether they be animal or vegetable'. What confused him especially were several types of molluscs which were 'fairly rooted, and in several cases perish if detached' – the rootedness was surely a plant trait, but other aspects of these molluscs looked more typically animal. There is understandable uncertainty as to which species Aristotle was referring to, although some type of mussel seems likely for one of his examples. He was apparently happy that sea-anemones were animals, but described sponges as 'in every respect like a vegetable'.[26] There are obvious non-microbial terrestrial examples too. Consider fungi – they don't look at all like animals but are not typically plant-like either. Theophrastus, a student and later colleague of Aristotle's, wrote extensively on botany and considered fungi to be plants, although again in many cases it's hard to be sure what species he was referring to. Indeed, in some cases there are disputes about identifications of some of his 'fungi', with alternative identifications including organisms as diverse as corals and true plants. However, he does seem to have written about truffles, and possibly puffball-like fungi too (Figure 6).

FIG 6. Common Earthball *Scleroderma citrinum* at the Lancashire Wildlife Trust reserve at Mere Sands Wood. Traditionally classed as plants, fungi look very different from most plants and formed an ambiguous group for naturalists trying to classify all organisms as either plant or animal. It seems likely that the ancient Greek philosopher Theophrastus referred to puffball- or earthball-like fungi in his writings – but, as Sharples & Minter (1983) pointed out, it would be unwise to try and identify the fungi mentioned by Theophrastus to species.

Microbes added all sorts of complications once people began to study them in detail. Van Leeuwenhoek considered what he was looking at were odd types of animals – 'animalcules'. However, while some moved like an animal, they were green in colour like a plant. Equally, some non-green ones were nevertheless apparently attached (rooted) to a substrate. If assigning them to either plant or animal, which character is most important, the lack of green or the rootedness? In many cases the microbes that seemed to move around at speed when viewed with a microscope were over time assigned to the zoologists for study, while the botanists acquired most of the others. A nice example of the common approach around the end of the nineteenth century is given by Eugenius Warming's *Handbook of Systematic Botany*, which was a gazetteer of all the main types of plants.[27] Warming was Professor of Botany in the University of Copenhagen, a leading European botanist of the time and a key figure in the founding of plant ecology as a subject. The first 119 pages of this 468-page book are largely devoted to organisms we would no longer consider plants (I am using the 1890 German edition, which was much more widely read than the original Danish version – an English edition came out a few years later). These include macroscopic organisms such as larger fungi, lichens and brown seaweeds, but also many microscopic organisms such as bacteria, diatoms, slime moulds and green algae; only these green algae would now fall within most biologists' definition of 'plants' (Figure 7).

While Warming was describing the general consensus of the time, during the nineteenth century various people had started suggesting that microbes were

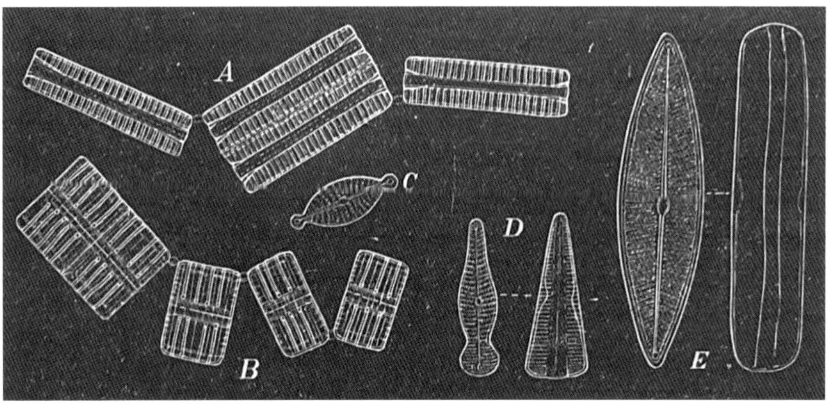

FIG 7. Diatoms illustrated by Warming (1890). These were considered plants by Warming and most other biologists in the late nineteenth century, but are now known to be only very distantly related to plants (see Figure 10). The illustrations show four different genera: *Diatoma* (A), *Tabellaria* (B), *Navicula* (C & E) and *Gomphonema* (D).

neither plant nor animal, and a new category was needed for them. In Britain in 1859 Richard Owen suggested calling this kingdom that was neither plant nor animal the Protozoa, while a year later John Hogg suggested they should be called the Protoctista (in the kingdom Regnum Primigenum). In the United States Thomas B. Wilson and John Cassin had suggested the term Primalia in 1863, while Ernst Haeckel, in the country that was just becoming Germany, suggested the word Protista in 1866.[28] The thing all these terms had in common is that they suggested these organisms were in some sense 'proto' or 'primary', coming at the base of any biological classification (Box 1).

Two of these terms have survived in use, albeit with modified definitions. Owen's Protozoa became a term for the phylum (or in some classifications a subkingdom or kingdom) including the predominantly non-photosynthetic eukaryotic microbes (see later in this chapter for a definition of eukaryotes). While this formal definition survived well into the second half of the twentieth century, it is now considered obsolete by many scientists, though there are still some dissenting voices.[31] However, Protozoa is still sometimes used as an

Box 1. Hierarchical classification in biology

The traditional approach to classification in biology is a hierarchical system, which dates back to the ideas of Linnaeus in the eighteenth century. Species are given a two-part name – such as the diatom *Navicula viridula*. The first word is the name of the genus and the second one the specific (species) name. A number of similar species will usually be grouped together in a single genus. Similar genera are grouped together into a family (in this case Naviculaceae). Families are grouped into classes and then classes into orders. These are grouped into a phylum. In this case the phylum is Bacillariophyta – which is made up of all the known diatoms, both living and fossil. There are at least 30,000 species of diatoms alive today, and probably more like 100,000.[29] Phyla are grouped into kingdoms, in this case the Chromista. Historically there were just two kingdoms, the plants and the animals. The number of kingdoms that should be recognised today is controversial, with six being the *smallest* number that has been argued for in recent decades.[30]

Kingdoms are now often grouped into domains, and three of these are currently recognised by most biologists (but see the complexities described in the main text). These are the Bacteria, Archaea and Eukarya (the eukaryotes). Diatoms are Eukarya, as are the majority of microbes covered in this book. As discussed in more detail in the main text, eukaryotes are now often grouped into a small number of 'supergroups', which don't neatly fit into the Linnaean hierarchy described in this box.

informal term for single-celled eukaryote microbes that are heterotrophic – that is, feeding on other organisms or chemicals produced by other organisms such as fragments of organic matter in soils or muds. This is an ecological definition, based on how they obtain their energy, not an evolutionary one, as many protozoa are only very distantly related to each other. I will use 'protozoa' in this sense in this book, as it's still a useful term in the context of a book that is mainly about the ecology of microbes (rather than about their classification and evolutionary history). Haeckel's term Protist is still widely used too. Until recently it was often used in biological classifications – as in the five-kingdoms classification discussed below. Now it's become an informal term for any eukaryotic organism that is not a plant, animal or fungus. Using it in this sense, the majority of organisms discussed in this book are protists. The term Primalia did not survive into more recent usage, and the same is mainly true for Protoctista – although the important and contrarian American biologist Lynn Margulis stubbornly preferred this term to the more commonly used Protist.[32]

Although during the nineteenth century there were multiple suggestions for a third kingdom to contain microbes and other organisms that didn't easily fit into either plants or animals, few biologists took these ideas seriously in the early twentieth century.[33] I suspect that part of the problem was that such ideas didn't easily fit into the structure of university departments or textbooks, which tended to be either botanical or zoological. It was not until the mid-twentieth century that an alternative to the plant/animal dichotomy started to become influential. However, even at the start of the 1960s a major introductory textbook on Protozoa could write that 'Some Protozoa are not animals, but plants'.[34] By the time I was an undergraduate student in the early 1980s a five-kingdom classification had become common and had made its way to the textbook I was using.[35]

This five-kingdom classification was developed by the plant ecologist Robert Whittaker from the late 1950s onwards. As an ecologist he was interested in what organisms do, not just in their evolutionary relationships. Because of this, his early attempts to break away from the simple plant/animal divide used three kingdoms: plants (the producers that sit at the base of food chains), animals (consumers that feed on plants or other animals), and fungi and bacteria (which form a decomposer kingdom, breaking down dead organisms). By the end of the 1960s he had expanded his classification to the influential five-kingdom approach, breaking up the decomposer group into separate fungal and bacterial kingdoms, and creating a protist kingdom for all non-bacterial microbes (plus a few larger organisms such as brown seaweeds which didn't neatly fit into the plant kingdom). This five-kingdom classification was enthusiastically taken up by

Lynn Margulis, who gave it a more evolutionary spin.[36] Although this system was very influential in the second half of the twentieth century, even with Margulis' vigorous defence it was undermined by large amounts of new data that became increasingly hard to fit into such a scheme, along with a growing consensus that evolutionary relationships – not ecology – should be the key arbiter of any biological classification. Much of the new data came from significant technological advances in biology from the mid-twentieth century onwards.

TECHNOLOGICAL DRIVERS OF CHANGE

It's not just technology that changes; the social behaviour of scientists changes over time too. Regular meetings in the pub, be it at lunchtime or after work, were far more common in the past than today. Perhaps the most famous such gathering in the history of biology was at lunchtime on 28 February 1953 in *The Eagle and Child* in Cambridge,[37] even though the story is largely apocryphal. As James Watson related the tale in *The Double Helix*, his 1968 autobiographical book on the discovery of the structure of DNA, Francis Crick arrived for his pub lunch loudly telling everyone within earshot that he had 'found the secret of life' – having with Watson just devised a workable model of DNA structure.[38] Years later, Crick told the science historian Robert Olby that he had no memory of such an event;[39] but dramatically the tale works beautifully to underline the importance of what Watson and Crick had done. That's rather the point, and even at the time of its publication Jacob Bronowski, in his review of *The Double Helix*, picked up on its archetypal structure, with key characters in the story being cast as the wicked witch, or the rival suitor – 'mythological postures rather than characters'.[40] That's perhaps why the book sold so well. It's a great story, based on fact, but somewhat modified in the retelling to improve the narrative.

Although molecular biology was beginning to become important in the mid-twentieth century, it would be some years before it started to have a large effect on microbial natural history. However, another technological advance around the same time quickly came to be influential in how scientists viewed the microworld. The basic idea of an electron microscope was developed in the 1930s, but it wasn't until a couple of decades later that working microscopes started to be more widely available: by the time of the First International Congress on Protozoology in 1961, talks on the results of electron microscopy 'were becoming commonplace'.[41] Electron microscopes use electrons rather than visible light to image the sample, and as the electrons have a much shorter wavelength these microscopes are able to resolve much more detail.

There are two basic types of electron microscope – transmission electron microscopes and scanning electron microscopes. A transmission electron microscope can be used to look at very thin sections through cells in extraordinary detail, while the scanning electron microscope allows you to look at the surface details of a cell, again at very high resolution. In Britain the Natural History Museum in London bought some of the first commercially available electron microscopes in 1965. The history of their acquisition comes from a very different time, with more readily available science funding and without the apparently endless grant applications often necessary to fund any expensive science today. The story is that Ron Hedley (then an assistant curator at the museum, but later its Director) saw electron microscopes demonstrated in an early edition of the BBC science and technology programme *Tomorrow's World*. Based on this, he contacted Cambridge Instruments and bought first a transmission electron microscope, followed soon after by an early scanning electron microscope.[42] One of the uses these microscopes were put to involved studies of testate amoebae by Hedley and Colin Ogden – who helped set up the electron microscope unit at the museum.[43]

Scanning electron microscopes can produce beautiful images of the organisms themselves, and it's these SEMs that are most relevant to a natural history book such as this one (Figure 8). However, it's what the transmission electron microscopes revealed about cell structure that is important to this introductory chapter. They clearly showed that the internal structure of bacterial cells is much

FIG 8. Scanning electron micrograph of a diatom frustule (the silica-rich cell wall – effectively a glass case). As they are created using electrons, rather than visible light, SEM images are black and white. This image has had colours added for effect. The species illustrated is *Actinophyta helioptera* from marine diatomite (diatom-rich sediments) in Maryland, USA. The diameter is around 250 μm (0.25 mm). This is a fossil diatom; because their cases are effectively glass, diatoms survive very well as fossils, and can be used to provide information about past environments when extracted from cores of marine or freshwater sediments. (Alex Hyde)

less complex than the cells of other organisms, including ourselves. Many of these differences could be seen with a good light microscope, but they became much clearer when looking at the results from transmission electron microscopy.

THE PROKARYOTE/EUKARYOTE DIVIDE

A key distinction in biology is that between the prokaryotes and the eukaryotes. These terms were coined by Édouard Chatton in some rather obscure publications in the 1930s. A few years later the Nobel-Prize-winning biologist André Lwoff started using them, but it was only in the early 1960s, when Lwoff introduced the Canadian bacteriologist Robert Yates Stanier to the terms during a sabbatical visit to Paris, that their use in biology took off.[44] In a lecture to the French Society of Microbiology on 2 March 1961, Stanier explained that the cells of bacteria and blue-green algae (now called cyanobacteria) are 'organised in a manner different from the cell one knows among the superior protists, among the plants and animals ... The cell type, which exists among the bacteria and blue green algae, is a procaryotic cell; the cell of the type which exists among the other organisms is a eucaryotic cell.'[45] Eukaryotic cells (the modern spelling normally uses a *k* rather than a *c*) contain a range of organelles – so called because of a perceived analogy to organs in an animal: as our heart, brain and kidneys perform different functions, so too do the organelles within a eukaryotic cell. So, for example, almost all the genetic material is found in the nucleus, the mitochondria are where energy is generated, and in eukaryotic algae photosynthesis happens in chloroplasts.

When I was an undergraduate student at the start of the 1980s, 20 years after Stanier's popularisation of these terms, my main general biology textbook considered these differences 'so fundamental that bacteria and blue-green algae are classified in a kingdom of their own ... and their cells are designated as procaryotic'.[46] However, even then things were starting to become more complicated – although my textbooks had yet to catch up with the changes. As the science historian Jan Sapp has pointed out, 'There was great confusion over kingdoms in the 1980s.'[47] The problem stemmed from attempts to apply the methods of molecular biology to an investigation of the evolutionary history of microbes.

Microbes in general, and bacteria in particular, had been especially problematic for people trying to reconstruct evolutionary relationships. The problem was that, compared to larger organisms, even with a good microscope there was limited morphological detail to use in such a reconstruction. Think of

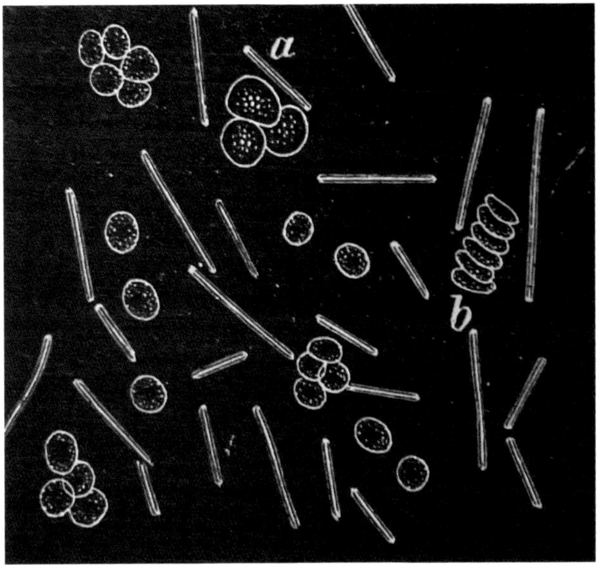

FIG 9. Late-nineteenth-century illustration of bacteria as viewed under a light microscope at high magnification (from Warming, 1890). The bacteria are just a collection of rods and spheres (cocci), with little morphological detail that could be of any use in trying to classify them.

the differences in structure that allow you to assign all the different cat species to one group, and differentiate them from all the species of dogs, from Red Fox *Vulpes vulpes* to Grey Wolf *Canis lupus*, or the many differences in flower structure that allow you to tell members of the pea family from the daisy family. In contrast to all this morphological diversity, bacteria are often just spheres or rod shapes (Figure 9), and it's hard to use such limited features to construct a natural classification (that is, a classification that is based on the organisms' evolutionary history).

By the mid-twentieth century Stanier was one of several leading microbiologists who had given up on the idea that a natural classification of prokaryotes would ever be feasible. However, in the late 1950s and early 1960s a few scientists – notably Francis Crick, Émile Zuckerkandl and Linus Pauling – had suggested that the new methods of molecular biology might allow scientists to construct 'family trees' of relatedness between different types of organisms. Building on these suggestions, during the 1970s Carl Woese and colleagues used molecular methods to look at short sequences of RNA. They chose to study a small part of the cell's ribosomal RNA (rather than the full complement of DNA) as this was more tractable with the technology of the time. All cells contain ribosomes, which are where proteins are made, and about half of all the material in the ribosome is RNA. The structure of the RNA is specified by the DNA, and as ribosomes are crucial to the functioning of the cells the guess was that the

RNA sequence in them would only change slowly over geological time, as any substantial mutation would mean that the cell could not function. This slow change made them a good candidate for investigating changes over long spans of geological time – so identifying relationships in the distant past. The techniques available to Woese at the time seem very primitive by today's standards – and a health and safety nightmare as well, using dangerous amounts of radioactive substances and highly flammable chemicals.[48]

The most famous results from this work came out late in 1977, and made it to the general news media, not just specialist scientific journals. The front page of the *New York Times* for 3 November 1977 claimed that 'scientists discover a way of life that predates higher organisms'. What Woese's group had shown was that the prokaryotes comprised two very different groups of organisms – not just bacteria but also another group that has come to be known as Archaea (at the time, they thought these were very 'primitive' organisms). Look in a university-level biology textbook today and you will often see life split into three basic types: Bacteria, Archaea and Eukarya (the eukaryotes). Given that this tripartite division of life became a textbook staple, and was front-page news for the *New York Times* when it was published, it seems odd that in a retrospective essay on this work Norman Pace, Jan Sapp and Nigel Goldenfeld could write that 'a fundamental breakthrough in biological science occurred in 1977, and most biologists did not notice.'[49] The lack of attention at the time was at least in part because the methods used in the study were so new and unfamiliar to most biologists that few knew what to make of it. In addition, it addressed questions that few biologists were studying at the time. The reason it appeared in *The New York Times* was because there had been a press release on the work. This is quite standard behaviour today but more unusual then, and some scientists felt that was not an appropriate way to publicise new science, undermining the credibility of the work.

How these three groups are related is an ongoing and complex question. The problem is that, especially in bacteria and archaea, genes can jump from one individual to another (so-called horizontal gene transfer – as seen, importantly, in genes for antibiotic resistance spreading from one type of bacterium to another). This complicates reconstructing evolutionary relationships of early organisms, as studying different genes may lead to you reconstructing different patterns of relationships. However, currently it looks most likely that the eukaryotes arose from *mainly* archaean ancestors; this means that in an evolutionary context the split between prokaryotes and eukaryotes is not fundamental. The crucial split is likely between bacteria (prokaryotes) and the archaea (also prokaryotes), from which the eukaryotes developed (albeit with inputs of bacterial genes too via horizontal gene transfer). However, the prokaryote/eukaryote divide does mark a structural difference between these two types of cells that still appears to be very important to many biologists. It's complicated, and shrouded in the mists of deep time, so quite how we should view the prokaryote/eukaryote split is still an open question.[50] However, as this book concentrates on eukaryotic microbes, and is primarily about their ecology rather than their evolutionary history, these details can be set aside in most of what follows.

It is only relatively recently that technology has developed enough to allow molecular methods to be used in studies of the classification of most protists, but these methods are often now the approach of choice. For example, testate amoebae provide many of the examples discussed in this book, and it was not until the mid-1990s that the first research papers applying molecular methods to testate amoebae started to appear,[51] but this approach has now greatly changed

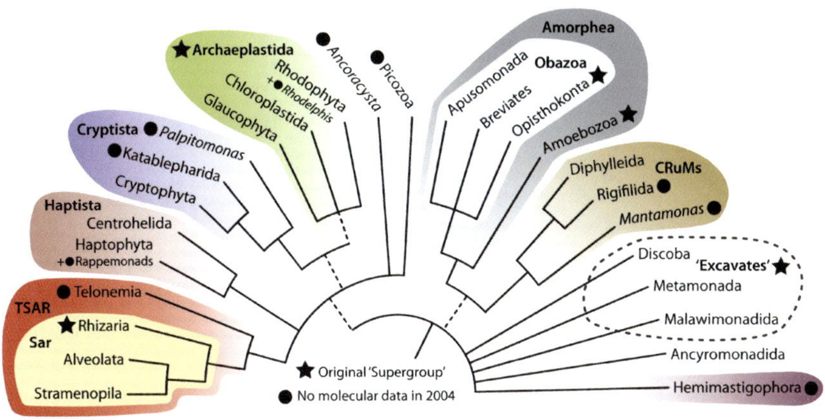

FIG 10. Eukaryotic supergroups, from Burki *et al.* (2020). This very high-level classification of the eukaryotes is based on molecular data. The animals and the fungi are in the Amorphea, while the true plants (along with the green and red algae) are in the Archaeplastida, and the brown seaweeds are in TSAR. Almost everything else is microscopic – so most eukaryotic diversity is microscopic. For example, it is currently estimated that around half of eukaryotic species are in TSAR, and with the exception of brown algae these are all microscopic, including all the diatoms and dinoflagellates. The testate amoebae are split between the Amorphea and TSAR, while the ciliates are also in the TSAR. The details of this supergroup classification are still evolving as more data become available; note that quite a few of the groups illustrated had no available molecular data in 2004 when the original attempt to construct a supergroup classification was published. (Reproduced from Burki *et al.*, 2020; Creative Commons licence, CC BY 4.0)

the classifications of many testates. As Anush Kosakyan and colleagues pointed out in 2016, this has led to a situation where the higher-level classification is uncertain, while at lower levels (e.g. genus) the taxonomy is constantly changing.[52] Indeed, many of the testate amoeba species have changed their names since I started to study them in the mid-1980s.

Advances in molecular biology, and our ability to compare organisms based on their genetic material, have also profoundly altered the highest levels of classification of eukaryotes over the last 20 years. Around the mid-2000s a new approach to their classification, based on molecular data, was developed that split all of the eukaryotes into around 5+ major groups – these became known as 'supergroups'. These groupings do not easily map onto the traditional higher-level groups in biological classifications; for example, both animals and fungi are included in a single supergroup. The testate amoebae, which are repeatedly used as examples in this book, are spread across two supergroups. This supergroup classification is continuing to be refined as more genetic data become available (Figure 10).

MOLECULAR METHODS AND ECOLOGY

Molecular methods have not just affected our views on the classification and evolutionary history of microbes, but have also influenced how we study their ecology. It is now possible to take an environmental sample, such as soil or water, and identify the types of microbes living in it using molecular methods, rather than time-consuming microscopy. Advances in the techniques of molecular biology, along with the developments of fast computers to handle the large amounts of data produced, have made such molecular approaches widely used by people interested in microbial ecology. A key technical development was the approach called 'massively parallel' or 'next generation' sequencing, around the start of the twenty-first century.

One way to think about how these new approaches fit into the longer history of microbial ecology is to consider the longstanding methodological difference between culture-dependent and culture-independent methods in studying the natural history of microbes living in the wild. Consider the larger organisms more familiar to most naturalists: if you want to census the numbers and types of birds on a saltmarsh, or plants in a meadow, you go out and count them. There may be all kinds of difficulties and complications, but in principle you can count the number of individuals of each species present, or at least estimate these data using a sampling approach. This is also the case with some groups of protists, especially ones that live in 'shells' or cases of some sort – such as diatoms and testate amoebae. These can be concentrated to give a density that makes counting practical (e.g. by sieving, or spinning the sample in a centrifuge), then they can be counted using a microscope. However, for many types, such as many protozoa including ciliates and flagellates, this approach is less easy, albeit often still possible. An alternative approach is to put a soil or water sample in a culture medium which supplies food for the protists. This allows them to increase in density to a point where you can more easily count them by microscopy. The obvious problem is that the relative proportions of the different species may no longer bear much relationship to their relative abundance in the wild (Figure 11). Also, some may not respond to the culture medium used and so be almost impossible to find on the microscope slide. Such culture-dependent approaches

FIG 11. Starlings *Sturnus vulgaris* at a winter roost at Attenborough Nature Reserve on the edge of Nottingham. It is clearly difficult to make an accurate count of the number of birds at such a roost, but it can be estimated (e.g. by counting birds on photographs such as this). It's clear that there are a large number of individuals. Contrast this with culture-dependent approaches to identifying protists. If you see large numbers of an individual on your microscope slide, is it because that species was actually common in the sample of soil or water you collected, or is it just because the culture conditions suited that species and something that was rare in your original sample is now very common on your microscope slide? This is often seen as a problem in much of microbial ecology, but is not an issue that usually confronts people studying large organisms such as Starlings.

may give you a partial species list (dominated by the species that respond to the culture conditions), but with limited information on numbers in the original sample. Molecular methods now give us an additional culture-independent way of enumerating the microbes in an environmental sample.[53]

The molecular approach to assessing microbial biodiversity that has become common since the start of the twenty-first century is metabarcoding. This looks for marker sequences, which are particular sequences of the chemicals (base pairs) that make up the genetic code. These allow you to list the diversity of genetic material present in a sample and then compare these with known databases of molecular sequences to identify which types of microbe are present. As it is looking for the range of genetic material in an environmental sample it's often referred to as an eDNA approach (the lowercase e standing for environmental). In practice, for microbes, it usually uses the sequences that describe the small subunit of ribosomal RNA (as used in the classic early work by Woese and colleagues described above). However, as with any method, this metabarcoding approach is not without problems. It involves a whole series of steps, any of which can potentially lead to biases in the results. A sample has to be collected from the wild and then undergo initial processing in the lab, before the genetic material is extracted and then amplified (that is, multiple copies are made to allow enough material for further analysis). There are then complex computing approaches to analyse these data and to compare them with existing databases. So there are multiple opportunities for 'errors and biases regarding both the relative abundance of organisms and the probability of their detection'.[54]

However, despite these potential difficulties, metabarcoding has revolutionised studies of the natural history of protists in recent years, and results from such studies will be discussed in several of the chapters in this book. But this book still makes extensive use of information from studies using microscopes, because there is a much longer history of such studies (and so a huge amount of relevant data), and it is also an approach more suitable for use at home, or in a school science laboratory. It's also often the case that using a mix of microscopy and molecular methods can tell us more about microbial natural history than using one method alone.

OUR KNOWLEDGE OF THE MICROBIAL WORLD IS VERY RECENT

The subject of this book is extraordinarily new. While a book on animals or plants could refer to studies made over 2,000 years ago, no one knew that the microworld existed until the microscope was invented in the seventeenth century, and it was not until the nineteenth century, and improvements in microscope design, that studies of microbes really took off. The strange collection of organisms that the microscope revealed put increasing strain on the idea that all of life could be classified as either plant or animal, and during the mid-twentieth century a variety of other potential classifications were suggested – with Whittaker's five-kingdom classification being the most influential at the time.

A fundamental structural division in biology is between prokaryotes (bacteria and archaea) and eukaryotes (with a more complex cell structure). In the mid-twentieth century this split was thought to mirror evolution, but in the 1970s the interpretation started to change. Results from molecular methods showed that the prokaryotes comprised two only distantly related groups, the true bacteria and the archaea, with archaea likely being more closely related to eukaryotes than to the bacteria.

This book concentrates on free-living microbial eukaryotes – the protists (there are also a few larger protists such as some seaweeds – see Chapter 7). As with the prokaryote/eukaryote divide, many scientists originally used the term protist in an evolutionary context; however, we now know that many protists are only distantly related to each other – for example, the red algae (which include some seaweeds as well as microscopic species) are more closely related to true plants than to most other protists. Molecular data are now commonly used to divide the eukaryotes into 5–8 supergroups – and the vast majority of organisms in these groups are microbial. This has an implication that is counterintuitive to many people. As all the prokaryotes are microbial, and the vast majority of eukaryotes are microbial too, this means that almost all of the diversity of life is too small to see without a microscope. Watch TV natural history programmes (or look at most books in this New Naturalist series), and you will get the impression that nearly all of biodiversity is made up of animals along with a few plants, and possibly the occasional fungus. This is completely wrong! My favourite, and so somewhat overused, way of emphasising this when giving talks is to say that to a first approximation David Attenborough's lifetime output of TV documentaries covers almost none of the diversity of life on Earth. If his output matched actual biodiversity, almost all of his documentaries would have been on microbes, and he would likely have made no programmes on vertebrates at all.

This almost unimaginable diversity of microorganisms creates a problem – how to fit an account of it into one book of just a few hundred pages. To keep things reasonably constrained, this book concentrates on free-living microbes (ignoring parasites), and largely ignores marine species too. It focuses on the eukaryotic microbes – the protists – with occasional discussions of prokaryotes (especially cyanobacteria, which are traditionally treated as 'algae', as ecologically they have much in common with eukaryotic algae).

This is an exciting time for anyone studying the natural history of microorganisms, as the technological changes of the last few decades have greatly increased our ability to study these tiny organisms. We have gone from first seeing them in the seventeenth century, to being able now to extract their DNA from environmental samples. Most of these technological improvements have happened since the mid-twentieth century. The microworld is now at one of the cutting edges of natural history. The eminent ecologist Edward O. Wilson – who specialised in ants – realised this and wrote the following on the final page of his autobiography, published in 1994:

> *If I could do it all over again, and relive my vision in the twenty-first century, I would be a microbial ecologist ... Into that world I would go with the aid of modern microscopy and molecular analysis.*[55]

As a first step on such a journey, the next chapter briefly outlines the microbial history of the planet, and also describes the microbes living today in some rather unpromising-looking habitats – such as hot and cold deserts.

CHAPTER 2

Microbial Planet

The ridge dominates the image in the viewfinder of my camera, slightly hazy as the day starts to warm up. It will turn out hot, at least by the standards of the north of Scotland, although snow patches still linger in the shadier gullies (Figure 12). The dramatic sculpturing of the ridge

FIG 12. An Teallach, in July 1983. 'The towers of our ridge curled like a whip-lash through the sky – flicked two miles around the corrie. An Teallach waited' (Murray, 1951, p. 205). The sediments that formed An Teallach's sandstones are thought to have started life as river sands (Brasier, 2009).

owes much to ice-age climates, but the rocks themselves are unimaginably old. These Torridonian sandstones were formed around a billion years ago. That's a thousand million years since the particles that have been cemented together to make these hard rocks were loose grains of sand. Some of the earliest-known non-marine protists have been found as fossils in these rocks where they are exposed on the coast to the north of where I stand. It's July 1983, I am nineteen, and in a year from now I will be starting to study Antarctic testate amoebae as a research project in the final year of my undergraduate degree at Coventry Polytechnic. The next few decades of my life will mix protists with mountains, and much more.

The main ridge of An Teallach consists of multiple summits comprising towers and pinnacles, all around 1,000 m above sea level. The mid-twentieth-century Scottish mountaineer W. H. (Bill) Murray was a great admirer of this mountain, where 'dark cliffs rise 500 metres to the sinuous summit-ridge, which in its middle section is a sharper knife-edge than any other on the mainland'.[1] Many of the pinnacles on the ridge can be bypassed – but not by a nineteen-year-old looking for the steepest scrambling in preference to the easier path. Memory is selective, and some of my old Kodachrome slides help fill the gaps from that day. I clearly remember stopping to watch and photograph Ptarmigan *Lagopus muta* high on the hill, but have no memory of the feral goat I also photographed. Although my memory is of having the hill completely to myself until I arrived at Shenavall Bothy that evening, one of my photographs shows distant figures silhouetted atop a rock pinnacle.

From the end of the main ridge I dropped down long scree slopes to the waters of Loch Toll an Lochain. I was now hot, and miraculously despite the lack of wind there were no midges, so with the corrie empty of people I stripped to cool off in the loch. While I have completely forgotten some parts of the day, such as the goat, I have a vivid memory of the lichens patterning the lochside rocks, and post-swim botanising along the shore looking at lichens as I dried off in the sun. At that time I don't think I could have put a name to a single species of crustose lichen – and indeed I still struggle with their identification, decades later – but their patterns made a striking impression, as much aesthetic as scientific.

I now think of lichens as an accessible bridge between the micro- and macroworlds. The main fungal body of the lichen is large enough to see with the naked eye, although often even more attractive when viewed with a hand lens, but much of its function – and its interest – is due to a range of symbiotic microbes living between the fungal hyphal threads. This makes lichens 'emerging macroscopic life forms, emanating from the interactions of microbial partners'.[2] More directly relevant to this chapter, it's likely that lichen-like organisms were among the early colonisers of the land, with the body of the fungus providing protection for its constituent microbes in this new and potentially hostile environment.

THE FIRST LIFE

Although these Torridonian sandstones are ancient, the rocks they sit on are vastly older. These are the rocks of the Lewisian complex – so named because they make up most of the Outer Hebrides, including the large island of Lewis. Many of them are double the age of the Torridonian rocks, a little over 2 billion years old. In the terminology of a geologist the sandstones sit unconformable on the Lewisian rocks. That is, there is a large time jump between the top of the Lewisian complex and the overlying Torrodonian sandstones; there are some rocks missing, and the sandstones were deposited on an old eroded rock surface. The Lewisian rocks are the oldest in Britain and look very similar to some very old rocks in Greenland. There is good reason for this, as it's thought that they were once connected, but then separated over the last 60 million years as the Atlantic Ocean formed.[3]

The connection with Greenland is interesting, as some of the earliest potential evidence of life we have comes from Greenland rocks. The Lewisian complex in the far northwest of Scotland is not a good place to go looking for fossils, however, as it is made up of metamorphic rocks. These are rocks that during their long history have spent time at great depth and been highly modified by the pressure and/or the internal heat of our planet. In Greenland in a few special places these rocks have survived in a somewhat less modified form, and it's here that people have claimed to find evidence for early life. However, for obvious reasons, the interpretation of these earliest traces of life has proved controversial. It involves looking for fossils of very simple microscopic cells, in rare rocks that have survived for several billion years. One of the problems is very similar to that faced by bacteriologists trying to classify bacteria before molecular methods became available, as described in the previous chapter. Many bacteria have extremely simple morphology (spheres or rods). When looking at a potential microfossil in a rock, how can you be sure that you are looking at a real fossil, rather than an artefact of non-biological processes that just happens to look spherical or rod-shaped? The same problem applies to potential microfossils in rocks from Mars. Nick Lane and Joana Xavier recently summarised the difficulties nicely in just one short sentence: 'The few geological traces that hint at early life are enigmatic.'[4]

What can we say about the very first microbes?[5] The Earth is around 4.6 billion years old. The oldest potential fossils from Greenland are microscopic carbon-rich spheres which could plausibly be the remains of bacterial-sized cells that have been affected, but not destroyed, by metamorphic processes. These are thought to be around 3.8 billion years old. This is interesting, but some way from being convincing proof of life. However, the chemistry of these carbon-rich structures is also intriguing. The nature of the carbon (technically, its isotopic makeup) is suggestive of carbon that has come from photosynthesis; if so, this is obviously really important, but the interpretation of these 'fossils' is still controversial.

Because of these problems, the oldest fossil microbes are often considered to come from the Apex Chert in Australia, and have been dated to around 3.5 billion years ago. These fossils look more convincing than the carbon-rich structures from Greenland but could still be non-biological artefacts in the view of some researchers. However, as with Greenland, there are chemical traces of life in these rocks too, and even scientists who are sceptical of the nature of the microfossils themselves concede that the chemistry implies a good chance of active biology at the time.[6] However, by 500 million years later, around 3 billion years ago, there is abundant evidence for life – with uncontroversial fossils such as stromatolites from rocks in several different locations around the world (Figure 13). So currently there is secure fossil evidence for life at 3 billion years, and good circumstantial evidence from isotope chemistry, along with less certain microfossils, for life between 3.8 and 3.5 billion.

FIG 13. Fossil stromatolites are laminated (sometimes domed) structures in rocks caused by the trapping and binding of sediment particles by microorganisms in aquatic systems. This example is around 2.6 billion years old and is from Coopers Cave, in the Cradle of Humankind (the site is also known for its fossils of human ancestors) northwest of Johannesburg in South Africa. It was most likely formed by photosynthetic cyanobacteria living in well-lit shallow waters.

There is however an alternative to fossils and isotope geochemistry for estimating the earliest date for life. It's possible, using a number of assumptions, to use the methods of molecular biology to estimate this based on the genetic relationships between modern organisms, and assumptions about the rate at which their DNA changes over time – an approach known as the molecular clock. Several recent studies using this approach have suggested dates of between 4.09 and 4.33 billion years for the ancestor of all of the life on the planet today.[7] These are very early dates that cannot be directly confirmed by fossils as there is no surviving rock of this age, so how you view these intriguingly early estimates depends on how you view the reliability of the molecular-clock approach.

How the first life evolved is another very difficult question. The two leading ideas are that it began around hydrothermal systems, most likely in deep water, with hot water and a wide range of chemicals emerging from the interior of the Earth, or that its origins lie in a 'soup' of chemicals produced in part by the effects of ultraviolet light – which would have to be in much shallower water. Charles Darwin speculated about an early version of this idea, writing of life starting in 'some warm little pond' in a letter to his friend J. D. Hooker in February 1871. Although these are the current best contenders for an explanation, there are many other suggestions, ranging from life arriving from elsewhere in the universe, to life starting inside droplets of immiscible liquids – that is, where two liquids remain separate and don't mix with each other. As Lane and Xavier have recently argued, it's likely too soon to aim for a consensus, as the 'question is too big'.[8] Developing several ideas to see where they lead looks more productive, an approach philosophers would call multiple working hypotheses.

All these very early fossils are thought to be of prokaryotes. However, this book is mainly about protists, eukaryotic microbes. When did these first evolve? As with the evidence for the earliest life, the fossils can be hard to interpret. Not only must we distinguish between microscopic fossils and similar-looking non-biological structures, but we also have to try and reliably tell fossil eukaryotes from prokaryotes.

THE FIRST PROTISTS

There are several different approaches to solving the riddle of when eukaryotic microbes started to appear. The most obvious is to look for microscopic fossils in the rocks. This usually involves making very thin sections of rock samples that can be viewed under a microscope, or chemically breaking down the rock matrix in the hope that any fossils will be left behind. As with the origin of life itself, you can also look for evidence of active biochemical processes in the rocks. In addition, the rise of molecular biology has led to molecular-clock approaches. As with the molecular attempts to estimate the date of the first life (or at least LUCA – the Last Universal Common Ancestor of all modern life on Earth), it's possible to estimate the date of the last common ancestor of all surviving eukaryotes (LECA – the Last Eukaryotic Common Ancestor). This is not necessarily the oldest eukaryote-like organism, as very plausibly some earlier lines will have gone extinct. Using conventional microfossils also has its problems – for example, it's very unlikely that the very earliest fossil of any given group, such as eukaryotes, will have both survived and been found by palaeontologists.

Because eukaryotic microbes (protists) have more structure than bacteria it's somewhat easier to recognise them as fossils – but it can still be far from straightforward. As I write, the earliest protist fossils that most people agree on come from sites in northern China and are around 1.65 billion years old (Figure 14). The current consensus for an estimate for LECA is slightly earlier at just over 1.75 billion. However, it's possible that eukaryotes evolved much earlier. If you take an optimistic (i.e. less critical) view of the evidence then it's possible to suggest an origin around 3.5 billion years ago.[9] The earliest protist fossils that many people are happy to assign to a known group are red algae from just over 1 billion years ago, although if you talk to experts in this area you will get a range of opinions on the reliability of the date of these fossils. The oldest 'protozoan' fossils (in this case

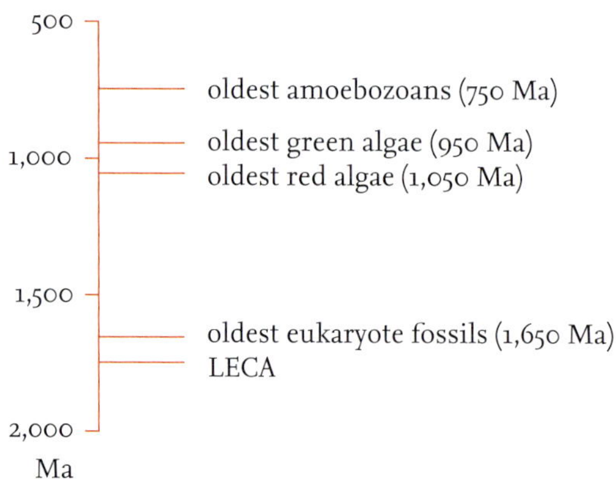

FIG 14. Summary timeline of early protist fossils (Ma, millions of years). LECA is the Last Eukaryotic Common Ancestor as estimated from molecular data – the current consensus is a date of just over 1,750 Ma, but the figure varies with different approaches to estimating its value. The oldest amoebozoans shown are the Grand Canyon testate amoebae. The Torridonian sandstones described in the text are around 1,000 Ma old. (Data from Miao *et al.*, 2024)

Amoebozoa) are of what appear to be testate amoebae from around 750 million years ago, found in rocks in the Grand Canyon, USA (Figure 15). Their shells make them more likely to survive as fossils than is the case for many protists.

These fossils from the Grand Canyon, and from a range of other rocks of a similar age, have been known since the 1930s and had been referred to by the non-committal term 'vase-shaped microfossils' or VSMs. It was not until the start of the twenty-first century that the Grand Canyon VSMs were identified as testate amoebae.[10] Certainly the photographs of these fossils look very like modern testates – to the extent that if you were shown them without being told their age you would in some cases be tempted to assign them to modern genera. For example, the Grand Canyon species *Palaeoarcella athanata* was given its generic name because it was difficult to distinguish it from testates in the modern genus *Arcella* (although recent research has tended to split the large genus *Arcella* into two different genera – see Figure 27, later in this chapter).

The study describing *P. athanata*, and eight other new species of fossil testates from the Grand Canyon, was written by Susannah Porter, Ralf Meisterfeld and

FIG 15. View north from Navajo Point on the south rim of the Grand Canyon, Arizona, USA. The testate amoebae fossil sites around Chuar Canyon are in the distance on the left-hand side of the river, in rocks that are 780–740 million years old. When Porter & Knoll (2000) first described these fossils as testate amoebae they thought it likely that they had fossils of both main groups (Arcellinida and Euglyphida), but it is now thought that all these fossils are likely arcellinids, and that the euglyphids evolved much later (Lahr *et al.*, 2019).

Andrew Knoll.[11] This is notable because at the time Ralf Meisterfeld was one of the leading global experts on the identification and classification of modern testate amoebae. An interesting question then arises: are these VSMs really very closely related to some modern testate amoebae, or is this an example of convergent evolution? That is, had a distantly related group of protists independently evolved shells that are almost identical to those of modern testates? This may be an almost impossible question to answer in a definitive way, and Porter and colleagues deliberately did not specify higher-level classifications for these VSMs, assigning them only genus and species names.

These testate-like VSMs are now described from a range of rocks from around the world, mostly dated around 730+ million years ago – but one site in south Kazakhstan may possibly be a little earlier at 800 million.[12] There are several fascinating aspects to these 730+-million-year-old 'testates'. Firstly, the environment they lived in is usually reconstructed as marine, although it is hard to rule out the possibility that they were terrestrial organisms washed down rivers into the marine sediments.[13] In the Grand Canyon example it's thought the rocks were laid down in a shallow bay connected to the wider ocean.[14] Modern testate amoebae are organisms of freshwaters, soils, or at most brackish systems – such as beach sands or saltmarsh. So this marine origin, if correct, appears to show a change in habitat over time. Perhaps even more interesting is that some of the VSMs have holes made in them, which are interpreted as evidence of attack by an unknown predator. This provides the oldest evidence of predation in the fossil record. Intriguingly, there are some modern protists, including the vampyrellid (vampire-like) amoebae, which feed in a very similar way.[15]

Potentially, predation is key to understanding much of the evolution of single-celled microbial protists. Susannah Porter and others have argued that the appearance of predation of protists by other protists around 740 million years ago helped drive the diversification of marine microbes at the time.[16] In addition, predation may also have played a role in the evolution of the first eukaryotic cells, although the details are controversial, with several rival interpretations available. As well as the presence of organelles, another thing that makes eukaryotes different from prokaryotes is the lack of a rigid cell wall. In eukaryotes such a cell wall is not universal, and the cell's shape is usually maintained by an internal collection of filaments and tubes, called the cytoskeleton (the analogy is to the skeleton in vertebrates that gives us support and shape). Because of their usually very rigid cell walls, most prokaryotes cannot engulf food particles from their environment, although a few archaea may be able to do this. Instead, prokaryotes absorb chemicals from outside their cell, and so have to release enzymes into the environment to externally digest their food before it is absorbed. In a few cases prokaryotes can penetrate the cell of their prey, and although this behaviour is sometimes described as 'predatory', parasitism seems a better term. The non-rigid cell wall seen in many protists allows them to take in food particles by a process known as phagocytosis. Before phagocytosis there could be no predation – if predatory behaviour is defined as actually *engulfing* (eating) all or part of the prey.[17] In addition, the cytoskeleton and lack of rigid cell walls were important in allowing sexual reproduction – for while bacteria can swap genes they don't have anything that looks like sex as a eukaryote would recognise it (see Chapter 6).

Phagocytosis is also potentially linked to the evolutionary origin of two important organelles in the eukaryote cell. People who have been taught much biology any time in the last 50 years may have at least a vague recollection that both mitochondria and chloroplasts are thought to have a symbiotic origin. Mitochondria are the site of energy production in eukaryotic cells, while chloroplasts are where photosynthesis takes place in photosynthetic eukaryotes (for more on the evolution of chloroplasts, see Chapter 4). Both of these organelles look a bit 'bacterial' when viewed with a microscope, and there were occasional heterodox suggestions, from as early as the late nineteenth century, that they may have originated as symbiotic bacteria living inside an early eukaryotic cell. But it was not until the late 1960s that this idea started to take off. The catalyst was a long paper on the origin of eukaryotic cells in the *Journal of Theoretical Biology*.[18] The young author was Lynn Sagan (she had been married to the astronomer Carl Sagan); she was then in her late twenties, but became better known as Lynn Margulis – Margulis being the name of her second husband (Figure 16). She has already appeared in this book as a key figure in the five-kingdom classification of life that was widely used in the late twentieth century. Indeed, the two are connected, as she saw her ideas

FIG 16. Lynn Margulis, with the microbiologist Ricardo Guerrero, at James Lovelock's 90th birthday party in 2009.

on the origin of eukaryotes as very much linked to her views on the five-kingdom classification.[19]

In her *Journal of Theoretical Biology* paper, Margulis argued that three organelles – mitochondria, chloroplasts and centrioles (these latter are structures involved in cell division) – had arisen through symbiosis. That is, they had originally been free-living bacteria which had been incorporated into what was to become the eukaryotic cell. A measure of how controversial this was at the time is the extreme difficulty she had in getting her paper accepted for publication in a scientific journal. Her paper was rejected multiple times – decades later, her memory was that it had around 'fifteen or so assorted rejections'.[20] However, the fact that it was an unusually long paper – 50 pages when finally published – would also have made it more difficult to publish.

It requires significant self-confidence and staying power to battle through so many rejections. I didn't get to meet Lynn until she was in her early sixties, but at that time she was certainly opinionated and very sure of herself, even when she seemed obviously wrong! However, she clearly loved her microbes, and had an encyclopaedic knowledge of their diversity. That level of stubbornness and inability to take criticism would often be very damaging in a scientist, but in the right person, at the right time, with the right idea, it can help change an entire field of science – and in the 1960s Lynn Margulis was just that person. She was also very charismatic, and it was obvious that even in her sixties her research students thought the world of her. This charisma seems to have given the story of her role in convincing other biologists of the origin of mitochondria and chloroplasts an almost mythic status. It's now textbook 'fact' that mitochondria and chloroplasts had such a symbiotic origin, although her idea that centrioles also had such an origin never got much support and is almost certainly wrong. As are her later suggestions that various other aspects of the eukaryote cell also evolved via symbiosis.

One of the things that made Lynn's *Journal of Theoretical Biology* paper stand out was its interdisciplinary nature, using lines of evidence not only from microbiology and cell biology, but from geology, atmospheric chemistry and genetics too.[21] In so doing she gave the symbiotic origin of some organelles a geological and ecological context – she very much had a naturalist's approach to microbiology. Also crucial was that by the time she was writing she was able to make use of early molecular data showing that small amounts of DNA had been found in mitochondria and chloroplasts – molecular genetic analysis was later to prove crucial in establishing that she was right to consider that mitochondria and chloroplasts had originally been free-living bacteria. In her 1967 paper she was not very explicit on how she thought these bacteria got into the proto-eukaryotic cell. However, she implied in passing that she thought this was through phagocytosis, and that the proto-eukaryote had therefore been trying to eat them. In the decades that followed, this certainly became the most commonly accepted scenario.[22] This idea became known as the 'serial endosymbiosis' theory, as the eukaryotic cell evolved in stages, acquiring phagocytosis which later allowed it to acquire the bacteria that gave rise to mitochondria, and later, in the case of photosynthetic protists, to acquire the ancestors of chloroplasts.

The serial endosymbiosis theory seems to be ecologically plausible, with the ancestors of mitochondria escaping being consumed by the cell, or possibly being deliberately kept within the cell, in effect 'farmed' for later use.[23] However, an alternative view is currently preferred by many, but not all, researchers interested in the topic. In this scenario two prokaryotes (an archaean and a bacterium) merge to form a eukaryotic cell in one event rather than in a series of steps.[24] This is still a symbiotic origin for the eukaryotic cell, but by a rather different mechanism. Certainly molecular studies suggest that LECA was a very modern-looking eukaryote with asexual cell division (mitosis), sex (meiosis), organelles and a non-bacterial cell membrane, which may favour the one-step origin idea.[25] Many modern prokaryotes are involved in so-called syntrophic interactions, where the combined efforts of two species are required to break down a food source. The suspicion is that at some point in the past two such organisms merged to give a eukaryotic cell. This seems extremely unlikely, but since the event appears to have happened only once we are presumably looking for just such an unlikely mechanism for the origin for eukaryotes.

THE FIRST TERRESTRIAL PROTISTS

The Torridonian sandstones that comprise An Teallach also outcrop at multiple places along the northwest coast of Scotland (Figure 17). At the very end of the nineteenth century the geologists Ben Peach and John Horne found phosphate-rich nodules in these rocks and sent them for further analysis, as Charles Darwin had speculated that such nodules might be a sign of early life, and in the nineteenth century no obvious fossils were known in these ancient rocks. Peach and Horne sent the rocks to Jephro Teall, who in 1901 became the Director of the Geological Survey, and was very active in studying the geology of northern Scotland. He thought he could see microfossils in these nodules when he examined them with a microscope. However, this potentially ground-breaking result attracted relatively little notice, as most people at the time considered such rocks too old for signs of life. They thought that Teall must have been looking

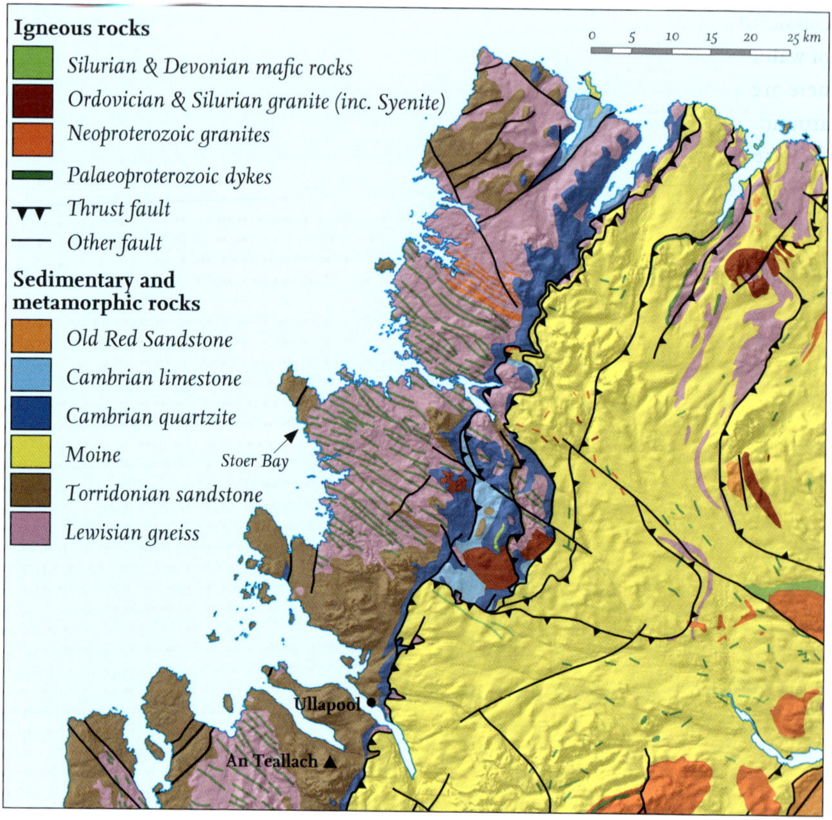

FIG 17. Geological map of northwest Scotland, showing the distribution of Torridonian sandstones and Lewisian complex (mainly gneiss, a coarse- to medium-grained metamorphic rock) along with the location of An Teallach and the important fossil site of Stoer Bay. The extensive Moine Supergroup rocks to the east are metamorphic rocks that were modified repeatedly from around 800 million years ago. (Modified from Friend, 2012)

at abiological structures that looked microbial by chance.[26] Half a century later, these fossils started to attract renewed attention, but it was the beginning of the twenty-first century before their full importance was realised.

Key sites for these fossils are Stoer Bay, and Enard Bay to the south, between Ullapool and Cape Wrath. This is a remote part of Scotland, a very long drive from anywhere. It must have felt even more remote to those nineteenth-century geologists. Some of the rock scenery is dramatic. Just north of Stoer Bay is the sea stack known as the Old Man of Stoer; first climbed in 1966, this is one of the classic adventure rock climbs in Britain, and even at low tide there are 8 metres of water to negotiate just to get to the base of the climb. The rocks along the coast here are all Torridonian, a mix of red mudstones, siltstones and sandstones, around 1 billion years old – although it's currently somewhat uncertain whether they are a bit over a billion, or slightly younger at around 990 million years. They also contain evidence for a meteorite impact, in the form of 'shocked' mineral grains preserved in the rocks.[27] Marine erosion along the coast has helped expose these rocks for geological study, as well as creating sea stacks such as the Old Man.

The fossils are of global significance for two reasons. Firstly, many are clearly eukaryotic – that is, they are microscopic protists. This is interesting, but, as described above, the first protists appeared at least 700 million years earlier – so these are a long way from being the earliest eukaryotes. Their real importance lies in their ecological context, as these Torridonian rocks are now interpreted as freshwater deposits, making them the earliest-known non-marine protists. The fossils from this remote part of Scotland are the first evidence we have of eukaryotic life on land. The question 'are these sediments marine or freshwater?' was still an open one in 1983 as I swam in Loch Toll an Lochain. It was not until the early 2000s that geological opinion moved firmly in favour of freshwater systems, albeit ones very different from the loch I swam in, being surrounded by a semi-arid landscape with no plants. Nothing like the much wetter, green landscape I looked across as I sat on lochside boulders warming up in the sun between swims in the cold water.

The evidence that these rocks are not marine deposits is multiple. Firstly, that unconformity between the much older Lewisian rocks and the Torridonian rocks is very informative. In several places it appears that the Torridonian rocks are lying on a weathered surface, possibly even a fossil soil. For this to be the case the surface of the Lewisian rocks must have been on land, and not under the sea. In addition to this are other lines of evidence for terrestrial conditions – such as sedimentary structures including cross-bedding, which is interpreted as being formed under river conditions. Additional lines of evidence come from the nature of the rocks themselves and their detailed chemistry.

Much of the twenty-first-century research that I have summarised above was led by Martin Brasier, in collaboration with various students and others, including his son Alexander, who is also a geologist.[28] The elder Brasier had a distinguished career studying the fossil history of early life on Earth, but sadly died in a car crash in 2014 just after he had retired from the University of Oxford. Martin Brasier and Lynn Margulis were friends, which led to Lynn spending 2009 as a visiting professor in Oxford, at a time when much of Brasier's work was focused on the Torridonian rocks. Alexander Brasier and colleagues described this period of Martin's 'intense discussion' with Lynn as 'a catalyst for some of his work on the origins of multicellularity' (a topic considered in Chapter 7).[29]

Although these Torridonian rocks provide the earliest evidence of eukaryotic life on land, this was clearly still aquatic – albeit freshwater rather than marine. It's thought that the earliest non-submerged terrestrial life was probably similar to the microbial biocrusts, sometimes called cryptogamic soils, which can be found in many arid habitats today.[30] Modern examples of these are described later in this chapter. It's possible that such crusts, made up of prokaryotes, may have appeared very early – between 3.8 and 2.5 billion years ago. However, the evidence is controversial and open to multiple interpretations, with much of the support for these very early crusts relying on inferring their existence from their potential effects on rock weathering.[31] For more familiar-looking terrestrial ecosystems with small plants we have to look in more recent rocks. One of the classic examples also comes from Scotland, not far from Aberdeen around the village of Rhynie.

THE RHYNIE CHERT

The silica-rich Rhynie chert is world-famous for preserving fossils of early land plants with an extraordinary level of detail – 'paleobotany's most iconic deposit'.[32] However, a visit to this iconic site will likely be disappointing, as there is nothing to see. The rocks don't outcrop on the surface, and so excavation has been required to collect samples. This raises an obvious question: how were the fossils discovered? They were first found by William Mackie, a local general practitioner and later public health doctor who conducted geological research in his spare time. The possibly apocryphal story is that while doing geological fieldwork near Rhynie in 1912 he sat down by a drystone wall for lunch and noticed small plant fossils in one of the wall's stones, which must have been quarried from a nearby site. It's a scene that would work beautifully if you were filming a historical reconstruction for a TV documentary, but it seems uncertain if it happened this way or if, less dramatically, Mackie noticed the fossils later in stones he had picked up quite literally in the field, by a wall.[33] The source of the fossils found by Mackie was traced by the Scottish Geological Survey to what the distinguished botanist Frederick Bower described as 'certain beds of chert of Old Red Sandstone age'.[34]

Writing in the 1930s, Bower used that phrase – 'Old Red Sandstone age' – because there was no way of knowing exact ages when he was writing, as dating using radioactive decay hadn't yet been perfected. We can now say that these

Devonian-Period rocks are around 407 million years old. The cherts, and the fossils within them, were formed in a system similar to the famous hot springs in Yellowstone National Park in the USA. Hydrothermal water rich in silica from time to time floods the surrounding vegetation. This not only preserves the external look of the plants (Figure 18) but can also preserve their internal structure, 'so that the tissues can be read almost as in a section of some living plant'.[35] This was of great interest to Bower, as much of his botanical research was based on using the structure and life histories of living plants to try and understand the evolution of early plants – as a Professor of Botany at Glasgow in the late nineteenth century he was 'a staunch Darwinist'. Fossils preserved in such detail provided an additional line of evidence to compare with his inferences drawn from modern-day plants, and much of the early study of the Rhynie fossils was carried out by younger colleagues of Bower.[36]

Although the Rhynie chert is most famous for its fossil plants, those silica-rich waters also preserved a range of microbes. Within the chert there are clearly preserved mycorrhizal fungi in some of the plant roots, which is important because it has been very plausibly suggested that mycorrhizae were key in helping plants to colonise terrestrial habitats. In addition, there is a lichen-like fossil, comprising fungus and cyanobacterium, although it looks rather different from modern lichens. There are protists too: in the soil around the plants were several types of algae, at least a dozen species of cyanobacteria and also microscopic

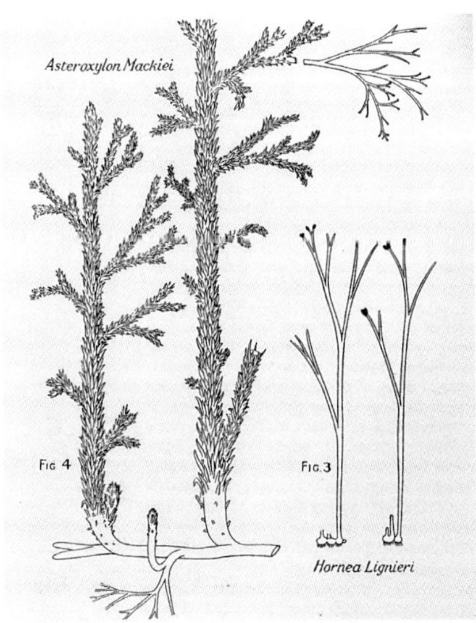

FIG 18. Two of the Rhynie plants from Bower (1935), based on reconstructions by R. W. Kidston and W. H. Lang, junior colleagues of Bower's who were responsible for much of the early work on these fossils. Note that *Hornea lignieri* is now called *Horneophyton lignieri* (Edwards *et al.*, 2017). The plants are drawn to the same scale, and *H. lignieri* was about 20 cm high. Edwards *et al.* (2017) provide a good short overview of the history of studies of the Rhynie plants. *H. lignieri* is one of the Rhynie plants in which fossils of mycorrhizal fungi have been found. So far three species of mycorrhizal fungi have been named from Rhynie plant fossils – two more, as yet un-named, are also known (Strullu-Derrien *et al.*, 2026).

green algae, along with a macroscopic alga in the form of a stonewort (see Chapter 4) called *Palaeonitella cranii*. There are also fossil microscopic water moulds (Oomycota), a group that used to be considered some type of 'primitive' fungus, but which are now known to be protists more closely related to diatoms and brown seaweeds. There are small animal fossils too, most interestingly fossil nematode 'worms'. Although nematodes are arguably the most numerous animals on Earth, they hardly ever fossilise, having no hard parts, and these Rhynie fossils are the earliest known.[37]

Although protozoa – that is, phagotrophic protists – are a common and important component of soil ecosystems, until recently none had been identified in the Rhynie chert. This is unsurprising, as like nematodes most have no hard parts and so are seldom fossilised. However, Christine Strullu-Derrien and colleagues have found tests which they convincingly argue are arcellinid testate amoebae (Figure 19). They have been found among cyanobacteria and green algae in what may have been a shallow pool, but one next to drier environments as it also contained the remains of the plant *Asteroxylon* – which is thought to have been terrestrial (Figure 18). Remember that the Grand Canyon testate amoebae (also thought to be arcellinids) were from apparently marine conditions. So the Rhynie chert suggests that both the ancestors of land plants and arcellinid testate amoebae may have moved from marine to fully terrestrial habitats via freshwater environments. The role of water is crucial in microworld ecology, and the rest of

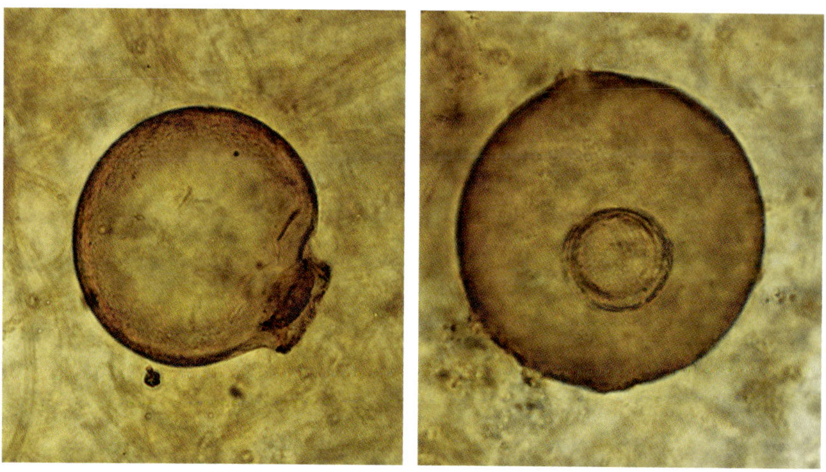

FIG 19. Two specimens of *Palaeoleptochlamys hassii*, fossil testate amoebae from the Rhynie chert (Strullu-Derrien *et al.*, 2019b). The tests are approximately 50 μm in diameter. (Christine Strullu-Derrien, courtesy of the Trustees of the Natural History Museum, London)

this chapter moves to the present day to consider extreme environments around the world where water may be in short supply – including biocrusts that likely resemble some of the first terrestrial communities.

THE IMPORTANCE OF WATER

The importance of water to protist natural history is obvious when considering marine or freshwater systems, be they contemporary or in the distant geological past. The importance of water to soil life is less obvious but still crucial, and one of the difficulties in life colonising the land must have been moving into an environment where water can be in short supply. At the start of the twentieth century protozoa in soils were thought to be ecologically uninteresting: they were seen as aquatic organisms, and when you saw them in soils it was because they had been washed in from nearby wet habitats. This idea started to change around 1910 when experiments at Rothamsted Agricultural Research Station, just north of London, showed that removing soil protozoa altered crop yield – suggesting that they had an important role in soil ecology and were not just uninteresting strays washed in from elsewhere.[38] Over the next decade or two the subject grew rapidly, and as Rothamsted biologist Harold Sandon wrote in the 1920s, in what was probably the first book to focus on the ecology and global distribution of soil protozoa:

> *The earlier objections based on the belief that the protozoa were present in the soil merely as stragglers brought in accidentally from other more natural habitats … have now been entirely removed … [and protozoa are] undoubtedly constituting an important component of soil population.*[39]

Water is however still important in soils. Soil is not a solid mass but contains many pores between the soil particles (see Chapter 8). In a very waterlogged soil these can be full of water, but often contain air with a thin film of water coating the surface of the soil particles. The more 'aquatic' soil organisms, be they microscopic animals (such as nematodes or tardigrades) or protists and bacteria, are restricted to the water-filled pores and/or water films coating the soil particles. When this water dries out, many forms can survive in an inactive form until the soil becomes damper again. Other very tiny animals verging on the microscopic, such as mites and springtails, use the drier air spaces in the pores.[40] These soil pores, their water content and microscopic inhabitants, are increasingly seen as important in understanding key aspects of the working of soil ecosystems. So although the nineteenth-century idea that protists are not important in soils because they are freshwater organisms that have been washed in from elsewhere is incorrect, they do still require some water in the soil. This might suggest that really arid soils should be lacking an interesting microworld – but protists and other microbes can survive, and even be ecologically important in such soils, often forming microbial crusts thought to be very similar to some of the earliest ecosystems on land. The final part of this chapter considers some examples of habitats where water can be in short supply, in both hot and cold environments around the world.

LIFE AT THE LIMITS: THE MICROWORLDS OF ARID HABITATS

Hot desert conditions are an obvious place to start when trying to get a feel for the extremes of life in the microworld. Here water is very clearly a limiting factor, as shown by the limited to non-existent cover of vegetation. Ascension Island in the South Atlantic is a remote volcanic tropical island, much of it covered by desert or semi-desert habitats (Figure 20). When Darwin visited in 1836, towards the end of the *Beagle* voyage, he described how the people of St Helena (the nearest other island) said 'we know we live on a rock, but the poor people of Ascension live on a cinder'.[41] However, even the most unpromising site – bare volcanic ash with only an occasional drought-adapted Ascension Island Spurge *Euphorbia origanoides* – had five species of protozoa when I was involved in studying them a couple of decades ago.

These protozoa must spend much of their time dormant, only becoming active under rare wetter conditions. In general, the protists of very arid soils are understudied, as the diversity and number of individuals is usually low, making them very time-consuming to study using microscopy. Many studies are small-scale, based on a relatively small number of samples; such as the Ascension Island study by Humphrey Smith and myself.[42] However, there are some more significant examples, such as the study of ciliates from 73 environmental samples from Namibia (in southwest Africa, and one of the world's driest countries). This was led by Wilhelm Foissner, and the results ran to 1,459 pages spread across two large volumes.[43] This study found 365 species of ciliate protozoa, of which 35 per cent were new to science – illustrating how little we know of protist diversity. Two of their samples were from next to *Welwitschia mirabilis* plants in the Namibian desert (Figure 21). These extraordinary and very long-lived gymnosperms have deep tap roots to reach water.[44] So from a ciliate's perspective the plant provides an oasis of organic matter (with associated bacteria to feed on), along with some

FIG 20. The view towards Sisters Peak on Ascension Island in the tropical South Atlantic (photo taken in 2003). The island is volcanic, and away from the highest summit (Green Mountain) it is very arid. There has been some tendency for the amount of vegetation to increase since this photograph was taken, possibly due to increasing rainfall with climate change. A soil sample from just below the summit of Sisters Peak (arid with very limited plant cover) contained three species of testate amoebae and nine species of 'naked' (i.e. non-testate) amoebae. A sample from the base of the Peak (on the other side from that shown) with a scattering of sparse vegetation produced a similar diversity. A sample from very arid conditions similar to that seen in the foreground of the photograph produced one species of testate and four species of naked amoebae – one ciliate and three flagellates (Wilkinson & Smith, 2006).

FIG 21. *Welwitschia mirabilis* growing in the Namibian desert. Soils from around these plants featured in the study by Foissner *et al.* (2002). Despite the unpromising desert habitat, they found up to 15 species of ciliate protozoa in these soils. (Tom Sherratt)

moisture, which may last for up to 1,000 years. Foissner and colleagues found 13–15 ciliate species in two samples associated with these plants, one of which was a new species. To put these numbers into context, in the same study in some savannah sites (by definition a mix of grasses and trees, mainly Mopane *Colophospermum mopane*) around 50 species of ciliates were found per sample, and in one case 126 species where ephemeral puddles were also sampled.

Of the Ascension Island samples studied by Humphrey Smith and me, the most species-rich was a biocrust sample – with 15 species of naked amoebae and 11 testate amoebae.[45] Biocrusts (also called cryptogamic soils or cryptogamic ground cover) are formed from a mix of mosses, lichens, cyanobacteria and other algae. A key theme of this book is that microbes are crucial components of all ecosystems; however, in most systems it's the larger organisms that are visually dominant. In these biocrusts – where the largest organisms are mosses and lichens – the importance of microorganisms is more obviously apparent. Such crusts are able to survive extremes of aridity and temperature and are important ground cover in many arid and semi-arid systems worldwide (Figures 22 and 23). They can also form in more temperate locations if plant cover has been restricted by disturbance or other factors, and in seasonally arid climates as seen in parts of the Mediterranean (Figure 24). These crusts can be very important in reducing soil erosion, and as already mentioned are thought to be similar to the earliest terrestrial communities.[46] They can also be important in the nitrogen cycle, as they contain cyanobacteria – which can fix nitrogen from the air, so increasing this important plant nutrient in the soil.

FIG 22. Desert habitat in Arches National Park, Utah, USA. Note the afternoon thunderstorm in the distance. The microbes of desert biocrusts (as seen in Figure 23) spend much of their life dried out and inactive but are adapted to make rapid use of water when available.

FIG 23. A biological soil crust in the Utah Desert, USA. In very dry conditions such as here the structure of the crust is often dominated by cyanobacteria. In slightly wetter, semi-arid conditions mosses and lichen become more important (as in the example from Crete in Figure 24).

Cyanobacteria can form an important part of the crust's structure, but can also be symbionts of crust lichens – while most lichens contain symbiotic green algae, some contain cyanobacteria as their photosynthetic symbiont. Importantly for the main topic of this book they also provide the habitat for a range of protists. For example, in a study of biocrust samples collected from desert soils at the Grand Canyon, USA, 51 species of ciliates, 28 naked amoebae and 17 testate amoebae species were identified. In addition, a small number of microscopic animals were found, namely several species of rotifers and nematodes.[47] While writing the first draft of this chapter I was also working in the lab looking at

FIG 24. Moss and lichen biocrust in garrigue vegetation at Paleochora in southwest Crete. The photograph was taken in April and there had been heavy overnight rain. This was likely some of the last significant rain before the very dry summer months when the mosses, lichens and microbes of the biocrust will become metabolically inactive (a topic discussed further in Chapter 6). Some of the more arid parts of the Mediterranean are good places for a European naturalist to see desert biocrusts.

moss-rich biocrusts from a semi-arid part of Victoria, Australia.[48] Several of the species I was finding are the same as those found in the Grand Canyon study – and also in the Ascension Island biocrust I looked at over 20 years ago. The testate amoebae in these crusts seem to be highly adaptable species that can respond quickly to wetter conditions, and have a global distribution. However, once molecular data are available it may turn out that there are cryptic species hiding among them. That is, species that, although genetically different, appear almost identical when viewed with a microscope (see Box 3 in Chapter 5).

Similar habitats to these desert biocrusts can be found in other dry conditions, such as lichens on rock surfaces, even in temperate climates. There are also many human-created habitats which can often be very dry and hot, and so have similar microbial communities. Good examples are mossy and/or lichen crusts on roofs (Figure 25). An interesting understudied urban environment consists of the cracks between the slabs of pavements – perhaps better referred to as sidewalks, as the limited studies have mainly been carried out in North America. Again, these can have much in common with desert biocrusts as they

FIG 25. Moss and lichens on a roof in Cley, Norfolk. These will dry out in hot dry weather but become active when moist, as will all the microorganisms living in them. In a summer heat wave the conditions on a roof like this will have much in common with a hot desert.

can become very hot and dry as the sun beats down on the sidewalk. A recent study of sidewalk microbes at Mississippi State University found a completely new genus of amoeba in the class Variosea – closely related to more familiar slime moulds.[49] As the authors of the study pointed out, this is a habitat that now has a near global distribution but hardly anyone has studied it. This is a nice example of Edward Wilson's point that to study the largely unknown world of microbial ecology he would 'need venture no farther than ten paces outside my laboratory building'.[50]

It is not only in hot conditions that water can be in short supply; in cold environments it may be unavailable because it is frozen. In March 1835 Charles Darwin was high in the Andes in South America, and on 'several patches of perpetual snow' he found a strange red colouration. This was 'red snow, so well-known from the accounts of Arctic navigators'.[51] At first he thought it was dust, but when he collected some, dried it on bits of paper, and later looked at it under a microscope he realised that it was not dust after all but a microorganism that was giving the snow its colour. Such red snow can be seen in persistent polar and mountain snow patches around the world. It is caused by algae – traditionally called *Chlamydomonas nivalis* (*Protococcus nivalis* in Darwin's time), but now known to be the product of a number of different green algae.[52] That's the first confusing thing about them: although red in pigmentation they are green algae (Chlorophyta), not red algae (Rhodophyta). The red colours are carotenoid pigments that help protect the algae from ultraviolet radiation at high altitudes and latitudes. As Darwin noted almost 200 years ago, they are found where the snow is partially melting so that liquid water is available for them to use.

If New Naturalist readers are familiar with red snow it's most likely from the Alps (Figure 26), but these algae can be found in persistent snow patches in Scotland too. However, there has been almost no serious research on Scottish red snow; the most detailed British study was published in the 1960s and runs to only two and a half pages describing algae from persistent snow patches in the Cairngorms.[53] At long last, as I write, a new study of Scottish snow algae using molecular methods has just started, based at the Scottish Association for Marine Science. Our lack of knowledge is worrying, as the habitat is becoming increasingly rare with climate change. Look back at the photograph of An Teallach that opens this chapter. It was taken in July 1983. The snow patches

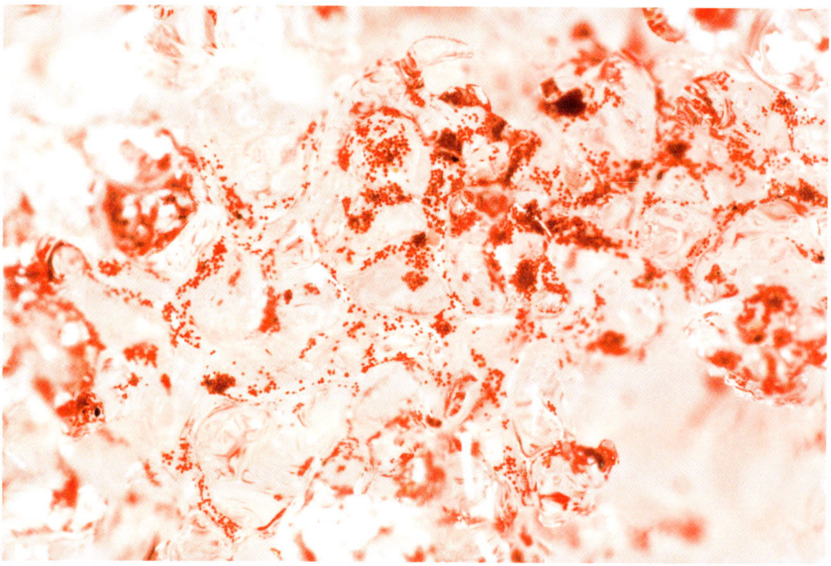

FIG 26. Red snow (due to algae) in July at 2,800 m in the Austrian Alps. One of the first scientists to suggest that red snows were caused by algae rather than dust was Robert Brown – probably best known today for giving his name to Brownian motion (see Chapter 3). In a list of plants he compiled from an early nineteenth-century Arctic voyage by Captain John Ross, he suggested the extensive red snow that had been seen at high latitude on the voyage was due to 'a type of very simple algae' (Brown, 1819). (Alex Hyde)

you can see would not have been there if I had tried to take it in 2023, for by the middle of June that year the snow patches on An Teallach had almost gone, and they vanished altogether in a heat wave over the next couple of weeks. Indeed, none of the persistent snow patches in the Scottish mountains survived over the summer of 2023.[54] Red snow habitat is rapidly declining in Scotland.

Antarctica is the extreme example of cold dry habitats, much more challenging for life than Alpine or Scottish snow patches. In some places temperatures can fall to lower than –80 °C, and less than 0.2 per cent of the land surface is ice-free. Much of the exposed land is on the Antarctic Peninsula – this is where much of the larger wildlife can live and so is where most ecotour cruise ships go. However, on the ice-free fragments of continental Antarctica, everywhere that biologists have looked they have found microbes and tiny invertebrates.[55] The cold makes it very arid, and this is arguably the most important feature affecting continental Antarctic soils as a habitat for life.[56] Some of the very dry, partially ice-free, valleys in the Antarctic seem so challenging for life that they have attracted the attention of biologists interested in the possibility of life on other planets – trying to answer the question 'how extreme can the environment be while allowing life to survive?' Recently Andrew Thompson and colleagues attempted to assemble a list of all terrestrial heterotrophic protists – that is, protozoa – recorded from across Antarctica.[57] Despite the challenging environment, which likely reduces the number of species, but also makes it logistically difficult to look for them, they recorded 236 types identified to species, and more examples identified only to higher taxonomic levels. These totals were based on data they had compiled from 54 different studies, from over 100 years of Antarctic research. For much of ice-free Antarctica, biodiversity is almost entirely microscopic, often associated with moss and lichen biocrusts (Figure 27).

MICROBES ARE EVERYWHERE

This chapter makes several very important points about the nature of the microworld. Firstly, for most of the history of life on Earth the only natural history was microbial, as there were no larger organisms. Even today much of biodiversity is microbial, and indeed this is a major theme of this book. The fact that microbes can survive in the very hottest, and coldest, deserts illustrates the ubiquitous nature of microbial diversity, and considering the microbes in these extreme environments draws attention to the need for water for active microbes – but also that many have the ability to wait in a dormant state until suitable conditions return.

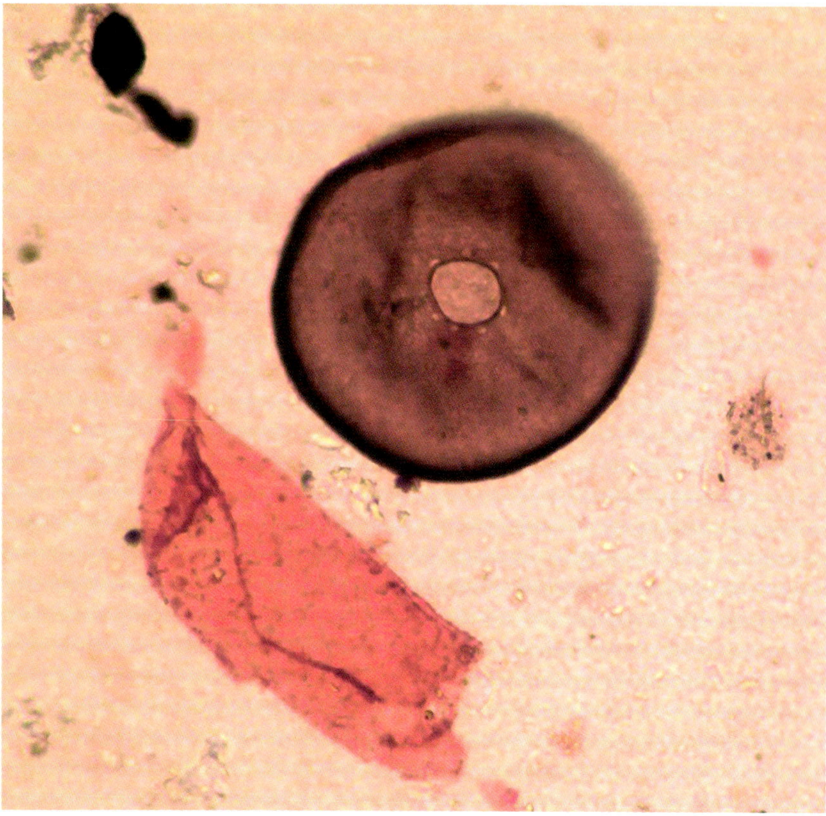

FIG 27. Testate amoeba *Galeripora arenaria* from a moss sample from the continental Antarctic, photographed from a slide prepared by the late Humphrey Smith as part of a study of continental Antarctic testate amoebae (Smith, 1992). This species used to be placed in the genus *Arcella*, which has recently been split (González-Miguéns *et al.*, 2022a), and it typically has a diameter of 60–130 μm. This morphospecies is widely distributed and is also found in Britain.

It is worth noting, however, that microscopic organisms experience the world very differently from large organisms like us. A microorganism 'swimming' in a Scottish loch would find water a very different, and more viscous, environment than would a human swimmer. Such differences form the subject of the next chapter.

CHAPTER 3

What is it Like to be Small?

Late afternoon on a hot day in early June and I am lying on the shingle at the Abbotsbury end of Chesil Beach in Dorset photographing Curled Dock *Rumex crispus* (Figure 28). It doesn't look like an easy place for a plant to grow – there is little to hold moisture in what passes for soil in the gaps between the pebbles, and the location is exposed to high winds and salt spray. This explains why the shingle is colonised by a relatively short list of (mainly specialist) plant species; the dock is unusual among these, as it is also found in a wide range of other habitats.

Although it needs an act of imagination for a human to view the environment in the way a rooted photosynthetic plant would, at least docks and people experience physics in a similar way. For example, in growing up into the air to

FIG 28. Curled Dock *Rumex crispus* growing at the western end of Chesil Beach, Dorset. A shingle beach is a difficult habitat for most plants to grow on. Many are specialists; the Curled Dock is one of the few characteristic shingle plants that are also found in a wide range of other habitats.

access the light it needs for photosynthesis, the dock has to work against gravity – just as in standing up after photographing it I too have to work against gravity. It's certainly a noticeable force when I pick up my heavy rucksack of camera equipment and field guides. That rucksack would feel lighter on a less massive planet like Mars, with lower gravity.

This intuitive feel for how other organisms experience their environment quickly breaks down when thinking about the inhabitants of the microworlds which form the subject of this book. While gravity is a significant force acting on both

FIG 29. British desmids in the genus *Penium* as illustrated by Edward Jenner in John Ralfs' classic monograph *The British Desmidieae* (1848). *Penium* are photosynthetic algae with a chloroplast in each half of the cell. They are particularly associated with acid water habitats (Bellinger & Sigee, 2015).

docks and humans, it's much less important to microorganisms. There are many other differences too. Compare a typical plant, such as the dock, with a microscopic photosynthetic alga such as a desmid (Figure 29). Plants tend to use leaves for photosynthesis, but algae don't have leaves, and those that do have rather leaf-like structures are macroscopic algae, the most familiar example being seaweeds. Why do some photosynthetic organisms need leaves, while others manage fine without?

This chapter describes the physical environment as experienced by microbes, a very counterintuitive world for an organism the size of a human. That means it's a chapter that might be hard going for human-sized readers, but it's worth the effort to understand a strange other world where gravity has very limited importance, water is thick and sticky, and molecules push you around in a way that makes moving in a straight line almost impossible if you are really tiny. As a way into this unfamiliar environment, I start by saying something about how the relationship between an organism's volume and surface area changes with size. It's a key bit of geometry which helps explain much about biology in general, and microbial natural history in particular.

SIZE, SURFACE AREA AND VOLUME

The way an organism's surface area changes in relation to its volume as its size changes is very important in understanding its natural history. When I was a student in the 1980s, John Maynard Smith's 1968 book *Mathematical Ideas in Biology* was influential in convincing me that relatively simple mathematics could tell you really interesting things about biology.[1] I still have my copy from that time, although the spine is now faded after decades of sitting on my bookshelves. Chapter 1 in that book is titled 'Some consequences of scale', and the chapter opens with the relationship between surface area and volume – both because it's a really important idea in biology, and also likely because the maths is simpler than that which comes later in the book. Think back to very basic school maths; one of the things you will have been taught early on is that the units of areas are given to the power 2 (e.g. cm^2) while volumes are to the power 3 (cm^3). The fact that the power associated with volume (3) is higher than that associated with surface area (2) means that as an organism gets larger its surface area increases more slowly than its volume. This idea is easy to grasp if you think of comparing two cubes (Figure 30). The larger cube, with edges that are three times longer than in the smaller version, will have nine times more surface area; however, its volume will be 27 times larger. Although few organisms are perfect cubes, considering such a simple hypothetical organism is a good way of grasping the basic idea. What follows from this simple maths is that a smaller organism has more surface area in relation to its volume than would be the case for a larger one of the same shape.

These ideas are often described in the context of explaining constraints on very large animals, for example the difficulty the largest mammals have in not overheating as they have a relatively small surface area (through which they can lose heat) compared to their volume (the tissue generating the heat).[2] Steven Vogel emphasised the point with his customary flare for a striking example:

> *If a cow used fuel at the same rate as a mouse, again relative to its body mass, its body temperature would exceed the boiling point of water, supplying us with precooked steaks.*[3]

However, this is a book about microbes, not mammals. A microbe's large surface area in relation to its volume means that it is much easier for it to take in all the chemicals it needs across the cell surface without having to invest in things like lungs or guts, which can be thought of as an animal's way of increasing its surface area for absorbing chemicals to match its large volume. This also applies to some borderline microscopic aquatic animals, such as flatworms (Platyhelminthes) or cnidarians such as *Hydra*, which do not need a circulatory system because their large surface-area-to-volume ratio means that diffusion through their surface is able to get oxygen to all their cells.

FIG 30. A simple illustration of the relationships between length, surface area and volume. A cube with edges that are three times longer will have nine times more surface area than the smaller version ($3^2 = 9$), but 27 times the volume ($3^3 = 27$). So the smaller cube has much more surface area in relation to its volume than the larger cube.

Leaves can also be viewed as a plant's way of increasing the surface area available for collecting light for photosynthesis. The earliest land plants didn't have leaves and presumably used green stems for photosynthesis; however, these plants were very small. One of the earliest plant fossils we have is *Cooksonia*, which was only around 1–2 cm tall and just 'the width of a needle',[4] a size and shape that would have given them a good surface-area-to-volume ratio for absorbing solar energy without having to resort to the solar panels we call leaves. The first *Cooksonia* fossils were found by W. H. Lang – who also made important early studies of the Rhynie chert plants and fungi described in the previous chapter. *Cooksonia* is not only smaller and more simple in structure than Rhynie chert plants (see Figure 18), but it is also a little earlier, from around 430 million years ago.[5]

The large surface area in relation to their volume seen in microscopic algae – such as the desmids illustrated in Figure 29 – means that they can collect all the light they need without having to increase their surface area by using leaves. A high surface-area-to-volume ratio has implications for the importance of gravity too, making a microbe's world very different from ours. As the slime mould expert John Tyler Bonner put it, 'there is a fundamental rift between the living worlds of the large and the very small.'[6]

SPLASHING HORSES AND FLOATING MICROBES

John Maynard Smith started out as an engineer, but after the Second World War he switched to biology and became a student of, and later a junior colleague of, J. B. S. Haldane at University College London. Haldane was a key figure in evolutionary theory and genetics in the first half of the twentieth century, and also wrote many science articles for a non-technical audience, which often incorporated original scientific ideas too. A frequently republished classic among these was his essay 'On being the right size'. In it he illustrated how gravity affects animals of different sizes, using the results of an imaginary experiment of dropping different-sized creatures down the shaft of a coal mine. He wrote:

> *You can drop a mouse down a thousand-yard mine shaft and, on arriving at the bottom, it gets a slight shock and walks away. A rat would probably be killed, though it can fall safely from the eleventh story of a building; a man is killed, a horse splashes.*[7]

In his textbook on *Comparative Biomechanics*, Steven Vogel related the story that many years ago the distinguished animal physiological ecologist Knut Schmidt-Nielson had expressed scepticism at this, so Vogel dropped two mice onto the pavement from the roof of a five-storey building to demonstrate the point: they were briefly stunned but otherwise uninjured.[8] Attitudes to animal experiments have changed over the years, and these days I cannot imagine any university research ethics committee being happy with dropping mice off the roof, even if it's unlikely to do them any serious harm!

There are several things contributing to the very different fates of Haldane's imaginary mouse and horse.[9] For a reasonably large organism (e.g. a human) falling some distance, such as down a mine shaft, much of the damage comes from the mass of the organism rather than the speed it is falling at. The words 'falling some distance' in the last sentence are important. What 'some distance' implies in this context is far enough for the falling organism to have reached terminal velocity. This is the speed where the downward force of gravity equals the upward force from the air resistance (technically, what a physicist would call drag). The force of gravity depends only on the mass of the object, whereas the air resistance increases with velocity – so as something falls it initially accelerates downward under gravity, but as the opposing force of air resistance gets stronger this acceleration decreases – eventually reaching zero at the terminal velocity. This is where the ideas of surface area to volume become important: they help save the mouse's life and give microbes the possibility of actually falling upwards.

Drag is related to surface area: the smaller you are, the more surface area you have in relation to your volume, and it's this that makes even a long freefall a trivial risk for the small. The microbe is experiencing a much greater upward force from air resistance than would be the case for a larger organism. This is how a simple parachute works – by greatly increasing your surface area you can safely jump from a plane. For most organisms, volume and mass are closely related. Mass can be defined as volume multiplied by density. The densities of organisms don't vary much with size: horses and mice are both made of the same stuff (bones, muscle, fat) in roughly the same proportions – so greater volume implies more mass. Small organisms combine low mass and high drag, both contributing to survival if they happen to fall down a mine shaft.

Falling under gravity matters in the real world, as well as in imaginary mine shafts. Look in an environmental physics textbook, and you will find equations for calculating the sedimentation velocity of a particle in a fluid such as water or air.[10] This allows you to calculate the rate at which a small particle, for example a microbe or a pollutant like soot, will sink through the air. For particles smaller than approximately 30 μm – which includes most microorganisms, along with many plant spores and pollen grains – a mathematical relationship called Stokes' law can be applied to calculate sedimentation velocity using particle size and density, along with a measure of the effect of gravity and the density and viscosity of air (or water). Even for somewhat larger particles, this approach would give an approximate measure of sedimentation velocity and avoid having to get involved with more complex calculations. If you look up the equation and do the calculations you will find that a spherical microbe (or pollen grain) with a diameter of 30 μm would sink through the air at 0.1 metres per second, while one with a diameter of 100 μm would fall at a faster rate of 1.2 metres per second.[11] As with the cubic microbe in Figure 30, considering the microbe as a simple shape, in this case a sphere, makes the maths more straightforward, allowing you to see the basic idea shorn of details and complexities. Such simple calculations can be done with pencil and paper, and with modern fast computers more realistic scenarios can be modelled too.

Such calculations are enough to illustrate the counterintuitive nature of falling for microorganisms. If we fall, be it down Haldane's mine shaft or off a cliff while botanising in the mountains, the outcome is obvious, painful and potentially fatal. However, if you are taking 10 seconds to fall 1 metre (as with the 30 μm-diameter sphere described above) then various options are possible. The air is seldom still, and at such low rates of sedimentation it's quite possible to fall upwards, as air movements take you up rather than down. You can't stay in updraughts for ever, and ultimately gravity will bring you down to Earth, but if you are wafted high into the atmosphere on strong air currents that may take a very long time. This has big implications for the dispersal of microbes around the globe, which will be discussed in more detail in Chapter 5. The same ideas apply to microbes in water – although here we usually talk of sinking rather than falling. Thus these ideas are also important to algae, and issues such as how quickly they may sink to depths where there is little light for photosynthesis. How microbes experience fluids such as water, and why they find them very different than we do, is the topic of the next section.

STICKY WATER: WHY MICROBES FIND IT HARD TO SWIM

Air is not the only fluid where microbes experience a very different environment to that which larger organisms are used to. If that sentence reads oddly to you, it's likely because your idea of fluid differs from that of a physicist. In physics, a fluid is any substance that responds to the application of a force by flowing. So the coffee in the mug next to me as I type this is clearly a fluid, as it will flow out of the way of a teaspoon placed in it. So is the air around me, as it flows out of my way as I move my hands to type. The keys on my computer keyboard are not fluid but solid: when I press them they move down under my finger rather than flowing round my finger out of the way.

So air and water are both fluids, but they feel different if you are moving through them. Think of fording a deep stream – it's harder to wade through water than to walk through air. This resistance of a fluid to you moving through it is effectively its viscosity. A more viscous fluid needs more effort to deform as it flows around a solid object – so it is harder work to push through a more viscous fluid. Air is less viscous than water, and water is less viscous than olive oil. This is a very informal way of thinking about viscosity, but it's good enough to get a feel for how an organism's size affects how it experiences moving through a fluid.[12]

An important aspect of viscosity is the so-called 'no-slip condition', although it's an odd and counterintuitive idea. Consider swimming in a lake – as in the opening vignette from the last chapter, or in the sea off Chesil Beach. When you get out of the water you are still wet, with a thin layer of water covering your skin. This is so familiar that we seldom stop to think, hang on, that's a bit odd! Why doesn't the water just fall off as you stand on the lake shore? Without a towel I dried off below An Teallach by sitting on rocks in the sun and wandering along the loch shore looking at plants and lichens, letting evaporation dry me before I got dressed. In fact, things are even more counterintuitive than this. While swimming, I was moving through the water, and yet the very thin layer of water that my body was in contact with was stationary, stuck to my body (the no-slip condition); it was the layers of water not in contact with my skin that slipped by as I moved through the loch. Although this seems very odd, the physics is clear – as shown by both theory and experimental results. This is how the world works.

Now think of what all this means for a microorganism moving through the water. That ultra-thin boundary layer of water on your skin makes little difference to your ability to swim, but its thickness is set by the size of water molecules (not by your size). So a small protist will have a water layer that is much thicker in relation to its size than you do. And the layer of water that it drags along with it may make a real difference to the microbe's ability to move through the water.

The way physicists try to make these ideas more rigorous is by using a dimensionless quantity called the Reynolds number. Here 'dimensionless' simply means that this number has no units (by units we mean things such as centimetres, or kilometres per hour). A look in an advanced physics text will reveal some complicated equations, but it's possible to get a feel for the nature of the Reynolds number by simply considering that it's proportional (signified by the symbol $\propto$) to inertial forces divided by viscous forces.[13] More formally:

$$\text{Reynolds number} \propto \text{Inertial forces} / \text{Viscous forces}$$

Inertial forces are those required to make an object change its speed or direction. The heavier it is, the more effort is required to alter its motion. Because of your mass, if you stop actively swimming your momentum will still carry you forward through the water – that is, you have substantial momentum and it takes time for drag to apply sufficient force to bring you to a halt. However, a tiny microbe has much less mass and so vastly less momentum. In summary, as objects moving in fluids get bigger, inertial forces dominate, whereas for small objects it is the viscous forces that are relatively stronger. This means that small organisms will have small Reynolds numbers, while large organisms have high Reynolds numbers (Table 1); even the largest microscopic organisms swim at Reynolds numbers less than one, and in most cases *much less* than one. So if you stop swimming you glide forward (inertial forces dominate), while a microbe stops dead the split second it stops swimming (viscosity dominates, and inertia has almost no effect).

Life at low Reynolds numbers (usually considered as any value under 10) seems odd to us. Not only do you stop moving abruptly if you stop actively swimming, but it also makes little difference if you are streamlined.[14] So there is no point in an aquatic microbe being shaped like a fish. Many protists move through water by beating lots of small hairs called cilia – this is where ciliate protozoa get their name – a strategy that works well at low Reynolds numbers but would not work for an organism the size of even a very small fish. Conversely, a microorganism could not swim like a fish. At low Reynolds numbers an organism trying to swim by, for example, moving a tail from side to side would make no progress, as in each cycle of tail wagging the water would return to its original

TABLE 1. Some approximate Reynolds numbers for organisms moving through a fluid (water or air). At low Reynolds numbers (< 10) the organism will find the fluid very 'sticky' to move through. Sources: Dusenbery, 2009; Vogel, 2013; Esteban & Fenchel, 2020.

	Organism	Reynolds number
In water	Bacterium swimming	0.0001
	Paramecium (ciliate protozoan)	0.075
	Rotifer (microscopic animal)	0.2
	Human swimming	1,000,000
	Large whale swimming	200,000,000
In air	Pollen grain or testate amoeba falling	0.01
	Small fly (e.g. fruit fly) flying	100
	Small bird flying	100,000

c.v
n
g
3
7
6
8
9
a
11
12
c.v
13
5
10
14
15
17
20
21
24
22
23
25
26
29
33
34
36
38
39
40
31
32
37
43
44
45
41

OPPOSITE: **FIG 31.** Flagellate protozoa from Volume 3 of Saville Kent's *Manual of the Infusoria* (1880–1882). Note the long flagella used for locomotion. As described in the text, these only work at low Reynolds numbers; a larger organism (swimming at a high Reynolds number) could not use flagella for propulsion. Traditionally the flagellate protozoa were considered to be all related to each other, usually being grouped together in the same class (e.g. Mackinnon & Hawes, 1961). However, it's now known that this flagellate appearance has evolved independently on several occasions – so flagellates are not monophyletic (i.e. they don't have a single common ancestor).

state. And the approaches to swimming used by many protists and bacteria could not work at high Reynolds numbers, as the cilia and flagella (thin hair- or whip-like structures; Figure 31) are so thin that they could not push enough water to move the organism by inertia, as with a tail-wagging fish.[15]

BROWNIAN MOTION: THE MICROBIAL ENVIRONMENT AT A MOLECULAR SCALE

One of the easiest ways to find protists is to look at a sample of water squeezed from a clump of wet moss (see the Appendix for a basic guide to looking for protists with a microscope). Chances are that as well as interesting microscopic organisms you will have some very tiny mineral particles in the water on your slide. Look at these tiny specks at a magnification of ×400 and they appear as tiny translucent particles, but you will also see something very odd. They are 'dancing' – although without any obvious choreography – jigging about in a random manner. Not moving in any particular direction, but certainly not staying still. Consider Figure 32, which shows part of a sample washed out of moss from the Cairngorms in Scotland. There is a large pollen grain, and alongside it the shell of a testate amoeba; both of these were motionless, but the smallest of the particles in the background were all moving about. This motion is called Brownian motion (after the nineteenth-century botanist Robert Brown), and for many years was very difficult to explain. However, it has important implications for the life of smaller microbes.

There is a good historical reason for including a pollen grain in the example I've used to introduce Brownian motion. Robert Brown became interested in what was to become Brownian motion after looking at pollen grains under the microscope and seeing small particles associated with the grains jiggling about. Despite what you will often read, it was not the pollen grains themselves that were moving, as these are too large to move in this way – although two usually reliable biophysics textbooks I consulted while writing this chapter repeated this error, as have many other writers.[16] Since Brown was looking at particles that had come from pollen, the

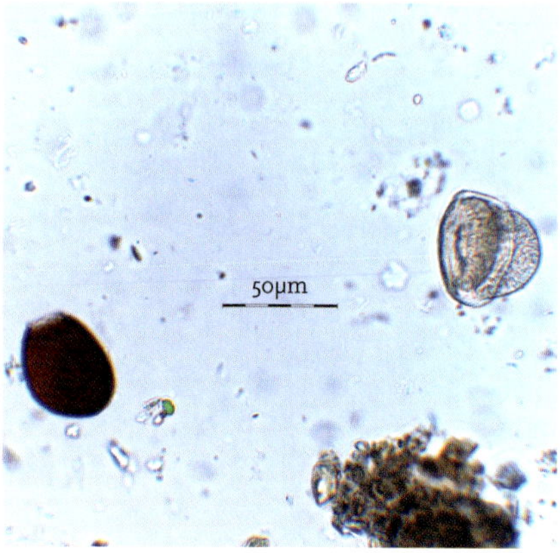

FIG 32. Sample from Cairngorm moss. To the right of the scale bar is a pine *Pinus* pollen grain (this is large by the standards of many pollen grains, and has air sacks to aid wind dispersal), while the darker object to the left of the scale bar is the shell of the testate amoeba *Assulina muscorum*. As can be seen by comparison with the 50 µm scale bar, both of these are around 50 µm (0.05 mm) long. These were stationary in this water-mounted sample when viewed under the microscope. However, the very smallest particles in the background – only a few micrometres (µm) across – were all jiggling about due to Brownian motion.

obvious thought was that they were moving because they were alive. However, he ruled this out by a series of simple, but effective, experiments. He looked at similar-sized particles that had come from fossil wood, and then particles that had never had any association with life at all – such as various minerals and metals. He still found them in motion, something which he could not explain but was confident was not associated with life. Brown's experiments were carried out at the end of the 1820s, but he didn't try to formally publish the results at the time, perhaps because he was lacking an explanation. Instead, he set them out in two short pamphlets – in 1828, and an update in 1829 – which he circulated to interested people (what today scientists would refer to as a 'preprint').[17] These were published after his death in the first volume of his collected research papers.[18]

Brown was not the first person to notice this movement of very small particles, but he felt he had given a much better description, and experimental investigation, of it than other 'discoverers'. He was reasonably justified in thinking this, but it is also the case in science that when several people describe the same thing around the same time, a discovery or idea often gets linked to the name of an already famous scientist such as Robert Brown. Indeed, this is so common that it has been named the 'Matthew effect', after the line in St Matthew's gospel, 'For unto every one that hath shall be given'.[19]

The explanation for Brownian motion had to wait for the start of the twentieth century, when over a few years several scientists developed a rigorous explanation for what is going on. In the spirit of the Matthew effect, it's worth pointing out that one of these was to become especially high-profile, namely Albert Einstein. Einstein did not at the time know about Robert Brown's observations, but in developing a mathematical understanding of the effect of atoms (whose existence was still somewhat controversial) he predicted in a 1905 research paper that you should be able to see their effect if you looked for them pushing around microscopic particles of the size Brown and others had been looking at. However, if you look in physics textbooks you will find a different mathematical treatment of Brownian motion, based on the work of Jean Baptiste Perrin and various collaborators, which was published a few years after Einstein's.[20] A good way to visualise what Einstein and Perrin were suggesting was proposed by Richard Feynman. He likened Brownian motion to watching a very large ball being thrown around by a group of people seen from a long way off, so only the ball is visible. However, you can infer the presence of the people from the movements of the ball as the invisible people interact with it, just as the invisible water molecules are interacting with the very small particles on a microscope slide.[21]

Brownian motion is very important in the world of the smallest microbes, that is ones with diameters of 2–3 μm or less. This covers many bacteria and the smallest eukaryotes, including some very small green algae. An aquatic organism with a diameter around 1 μm will be constantly jiggled around by water molecules, unless it is attached to a surface. Such an organism therefore cannot maintain any particular orientation within the waterbody and so cannot easily identify and move towards a food source, or similarly identify and move away from trouble. If you have ever found yourself in a large crowd struggling to walk in a straight line as you are being pushed around, you may get a feel for the microbe's lack of control of its own movement and orientation.

Although Brownian motion restricts a very small microbe's ability to act in a way that intuitively seems sensible to humans – namely, to identify something in the environment and move towards (or away from) it – it can aid an organism in moving through its environment in search of resources in a way that seems very odd to us.[22] Such a microbe will have little control over its movements, but Brownian motion allows it to move around by a process known as a 'random walk'. Although on average you would expect the organism to not move, as random collisions with water molecules will be just as likely to move it one way or another, in practice these random processes will allow an individual organism to move over time. Visualising things in three dimensions can be difficult, but the basic idea can be easily grasped by thinking about a one-dimensional example – where an organism can move either left or right over time (Figure 33). The organism will tend by chance to drift away from its starting point, potentially allowing it to find new food sources, but by chance rather than through active searching. This will apply only to the smallest protists, but it does apply to many bacteria – which can be brought within capture range of predatory protists by the random walks of Brownian motion.

MICROBES EXPERIENCE A VERY DIFFERENT WORLD FROM HUMANS

The physics of the microworld seems very odd to a human-sized organism. Equally, if a very small microbe could think about what it is like to be human, it would consider it extremely strange that falling from a height would likely kill you, that you can continue to glide through water when you have stopped actively swimming, or that you can purposely move from A to B without being pushed around by the effects of Brownian motion. Such a microbe would need a lot of imagination, and a good grasp of physics, to get a feel for what it's like to be a large organism. For example, it would not be immediately obvious to it why most fish have streamlined shapes, just as it seems common sense to us that a microbe that swims through water would benefit from being streamlined, although this is in fact wrong. The fact that microbes experience the physical environment differently from the large organisms that most naturalists and ecologists study raises the possibility that their ecology may turn out to be very different from what we are used to from studying insects, birds or flowering plants.[23]

Understanding the way microbes experience the physical world explains why microscopic algae don't need leaves, why microbes may be able to easily travel intercontinental distances on air currents, and why many use cilia or flagella to power their movements – something we don't see in the more familiar larger organisms. Several of these ideas will be made use of in the following chapter.

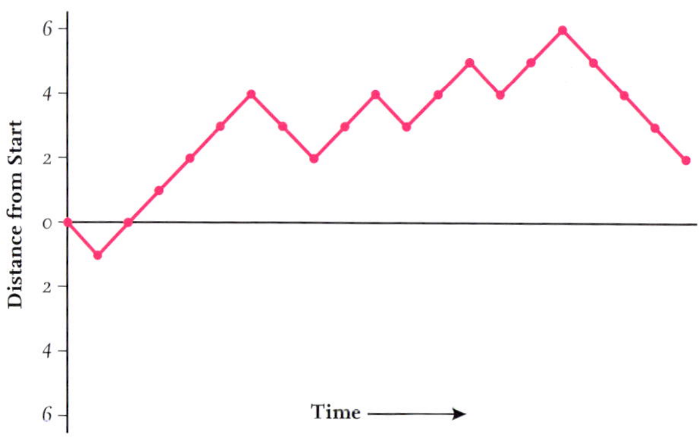

FIG 33. An illustration of a random walk in one dimension. The graph shows the results of 20 coin tosses where heads leads to an upward movement and tails to a downward one. In this case, after 20 time steps (20 coin tosses) the organism has moved two units away from where it started. However, over the course of those 20 time steps it has moved quite some distance, potentially finding food – or other resources – during its travels. For a microbe experiencing a Brownian motion random walk this would be happening in three dimensions, but the idea is more easily grasped (and graphed) in this simple model with one spatial dimension.

CHAPTER 4

Making a Living: Algae and Protozoa

Cwm Idwal in north Wales: it's a sunny day in late June and in quiet moments there is only the noise of distant streams and an occasional sheep. At other times the peace is temporarily broken by the noisy chatter of school geography field trips, although most of these seem to only make it to the edge of the Cwm. By the shore of Llyn Idwal, tiny insects dance in the afternoon sunlight as I take photographs along the lake edge, while a teacher attempts to round up his charges on the slope behind me.

The school groups are here because this is one of the classic locations of British physical geography. In 1954, Cwm Idwal became the first National Nature Reserve in Wales because of its geological interest, and also because it's a classic site for mountain plants.[1] However, I am here looking for algae – and while the Llyn seems an obvious habitat for such organisms, there are algae away from the lake too, and some of these have the advantage that they can be found without the aid of a microscope, as they are coating the surfaces of damp rocks. While the individual cells are too small to see, the microbes in quantity are visible to the naked eye.

I hike further into the Cwm. It's early afternoon, with the light catching the Idwal slabs – great sweeps of off-vertical rocks which have been a feature of north Wales rock climbing since the late nineteenth century (Figure 34). In the mid-1960s Timothy Allen also visited these slabs as part of his PhD work at what is now Bangor University. He collected algae growing on rock surfaces at 74 sites spread across six different locations around the Cwm Idwal area over four days of fieldwork in September 1966, including from rocks around the slabs. His main focus was on investigating the utility of multivariate statistics on studying algal communities, an approach which at that time had been little used on algae but was starting to become important as computers were making such complex calculations more practicable. In the research paper he wrote on this work he listed 12 species of algae as 'predominant' on Cwm Idwal rock surfaces – seven cyanobacteria and five green algae.[2]

Algae need water, and just to the side of the main slabs is a large wet sloping rock. The surface is covered with a dark slimy gunge, shining in the early-summer sunlight (Figure 35). Although accurate, that's a rather unflattering description of something so important and interesting. With a hand lens it's obvious that this gunge is alive – living photosynthetic slime. It's a microbial biofilm, and the colours suggest that many of the organisms that comprise it are cyanobacteria, probably some of the same species of algae recorded over 50 years ago by Allen. Such slippery biofilms have been causing trouble for rock climbers since the start of the sport, and are one of the reasons that climbers usually wait for a heatwave before attempting anything difficult on the high mountain crags.

FIG 34. Cwm Idwal, with the Idwal slabs sunlit right of centre, above the southern end of Llyn Idwal. The wet boulder shown in Figure 35 is just above a path to the left of the slabs. The Cwm was highlighted as containing glacial geological features deserving of national-level protection in the 1947 government report on *Conservation of Nature in England and Wales* (the committee that produced the report was initially chaired by Julian Huxley, one of the founding editors of the New Naturalist series). This committee's report played a key role in the founding of National Nature Reserves, of which Cwm Idwal was the first to be established in Wales. I described other aspects of the Cwm's natural history, especially the vegetation, in my earlier New Naturalist *Ecology and Natural History* (Wilkinson, 2021). The photograph was taken in June 2024, as were all the other Cwm Idwal photos in this chapter.

FIG 35. Wet rock with microbial biofilm, and the Idwal slabs in the background. This dark-coloured biofilm likely contains cyanobacteria.

These are not the only microbial biofilms to be found in Cwm Idwal. Down by the Llyn some of the rocks just above the waterline are stained black, something you can see widely along the margins of similar nutrient-poor lakes around the western mountains of Britain and Ireland (Figure 36). That's what I had been photographing earlier in the day while small clouds of insects caught the sunlight. This staining is caused by encrusting cyanobacteria. In looking at microbial films such as these (Figures 35 and 36) you can consider yourself a time traveller, as you are looking at something that is probably similar to some of the earliest life on land (see Chapter 2).

There are other algae in the Cwm that you can see without resorting to microscopy. In a stream, close to the wet microbial biofilm I have been looking at, the rocks are covered in filamentous green algae (Figure 37). Allen recorded similar algae on Cwm Idwal rocks over half a century ago – calling them *Zygnema ericetorum* and *Zygogonium* sp. The correct classification of these algae is still an area of ongoing research, and it's not really possible to be sure which species Allen saw. The same goes for me: because I was in a National Nature Reserve I didn't collect any samples for microscopic analysis, so can only say that the algae I saw were of that general filamentous type. As Stancheva and colleagues wrote a few years ago, 'The enigmatic green algal genus *Zygogonium* Kützing and its type species *Zygogonium ericetorum* have confounded phycologists for decades.'[3] (Phycology is the technical term for the study of algae – so people who study algae are phycologists.)

This illustrates an important problem in studying microbial natural history: there have been very substantial name changes for many of these organisms, especially in recent decades – such as the change in genus from *Zygnema* to

FIG 36. Small rocky island in Llyn Idwal – note the vegetation, which has been protected from the historic high levels of sheep grazing (see Wilkinson, 2021). This grazing is now much reduced (I only saw four sheep in the Cwm on the day I took this photo). More directly relevant to the themes of this book is the band of black on the rock just above the waterline. This is commonly seen on rocks in the nutrient-poor mountain lakes of western Britain and Ireland, and comprises mainly cyanobacteria, along with some lichens (e.g. the cyanolichen *Ephebe lanata*). The cyanobacteria are usually principally in the genera *Gloeocapsa* and *Stogonema* (Pentecost & Whitton, 2012), both genera recorded in Cwm Idwal by Timothy Allen in the 1960s.

FIG 37. Filamentous green algae in a Cwm Idwal stream. Probably similar to the algae that Allen (1971) called *Zygnema ericetorum*, but the taxonomy of these algae is rather confused.

Zygogonium for *Z. ericetorum*. A trivial, if memorable, example is the green alga *Hormidium flaccidum*, also recorded on Cwm Idwal rocks by Allen. Just after Allen published it was realised that *Hormidium* had already been used as the name for a genus of orchids, and so was not available to use for algae, and the name was changed to *Klebsormidium flaccidum*.[4] However, most of the name changes have been driven by our rapidly developing understanding of the evolutionary relationships of various microbial groups, often due to the use of molecular methods.

Although the wetter rock surfaces and boulders in streams are an accessible place to look for algae, the most obvious habitat is Llyn Idwal itself. The lake is well known for its aquatic plants, especially the so-called isoetid flora made up of small plants with rosettes of cylindrical-looking leaves – such as Quillwort *Isoetes lacustris*, Water Lobelia *Lobelia dortmanna* and Shoreweed *Littorella uniflora*. Taking off my boots to paddle and look for these plants, I could feel an algal biofilm, slippery under my feet. There are larger algae here too masquerading as vascular plants. Several species of stonewort *Nitella* have been recorded from the lake over the years, although some of the species listed for this location in the past may have been misidentified.[5]

Stoneworts are large multicellular green algae that can be 10–20 cm tall and look like a conventional plant (Figure 38). Effectively they are a freshwater

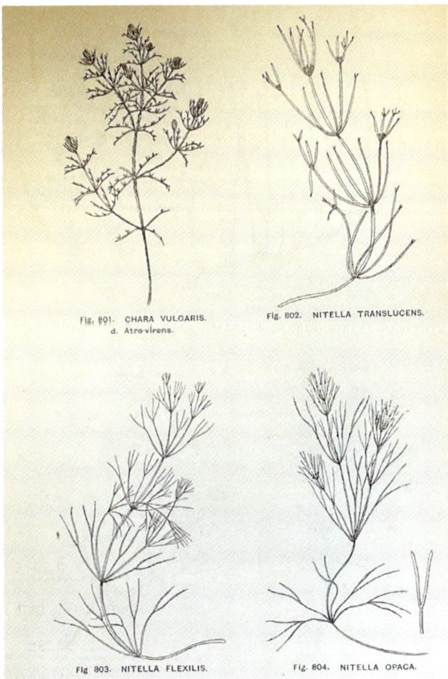

FIG 38. Stoneworts: one species in the genus *Chara* and three in *Nitella* illustrated by Emily Margaret Wood from the 1902 *Flora of the Liverpool District* (Green, 1902). Species of *Chara* live in more alkaline lakes and are often hard to the touch because of encrustations of lime (Bellinger & Sigee, 2015). This explains the group's common name of 'stoneworts'. *Nitella* can live in more acidic water and does not feel stony. Although they are green algae, stoneworts look remarkably like conventional vascular plants. This helps explain the strange historical quirk that the plants that form the area of interest of the Botanical Society of Britain and Ireland are vascular plants *plus* stoneworts (the other algae, along with bryophytes etc., fall outside its remit and are claimed by other societies, leaving Britain and Ireland with a rather oddly focused botanical society).

equivalent to seaweeds – which are also large algae the size of more familiar plants. All living examples of stoneworts are classified in a single order, the Charales, and have already been mentioned in Chapter 2, as fossils of them have been found in the Rhynie chert. The examples I found in Llyn Idwal seemed relatively uncommon despite them being described as 'abundant' in this part of the lake in the late twentieth century. They were also covered in epiphytic algae, which may be a sign of some nutrient enrichment in the lake (Figure 39). Pollution has been making much of Britain more nutrient-rich in recent decades, but in the case of Cwm Idwal there is also the suspicion that changes in the vegetation caused by greatly reduced grazing levels may have led to more nitrogen compounds leaching from the soils into the lake.[6]

There are microscopic algae in these waters too. The earliest described were various desmids (a group of green algae) in the late nineteenth century. Indeed, by the start of the twentieth century it had become clear that the mountain lakes of the west of Britain were particularly rich in desmids.[7] There are plenty of species of diatoms as well, many of them growing on the rocks and sediment in Llyn Idwal – the commonest being *Achnanthes minutissima*.[8] These diatoms will have been contributing to the slippery biofilm under my feet as I paddled in search of aquatic plants. In addition to all these photosynthetic algae there will be heterotrophic protists – protozoa feeding on the algae and other microbes. The rest of this chapter introduces these various ways of making a living, by photosynthesis, heterotrophy or mixotrophy (a mix of the two). In doing this I also give an overview of the diversity of free-living protists – albeit with an ecological rather than a taxonomic focus.

FIG 39. Stonewort *Nitella* sp. in shallow water in Llyn Idwal, June 2024. Note that it is covered with epiphytic algae (an epiphyte is a non-parasitic plant that grows on another plant).

THE NATURE OF ALGAE

Most of the algae I described from Cwm Idwal were either green algae or cyanobacteria, although diatoms got a brief mention too. The first difficulty in writing about such organisms is to decide what to call them. Green algae and cyanobacteria (called blue-green algae in older books) are only very distantly related, as cyanobacteria are prokaryotes while green algae are true plants (in the kingdom Plantae, sometimes now referred to as the Archaeplastida, as in Figure 10). The red algae are also usually now included in the kingdom Plantae, albeit in a separate subkingdom from the green algae and the more familiar plants such as mosses and flowering plants. Diatoms, on the other hand, are considered very distantly related to the true plants, but more closely related to brown seaweeds.

This confusing picture makes many biologists uncomfortable with the term algae, as its members are only very distantly related to each other. However, as John Bolton has argued, non-taxonomic terms for groups of organisms can be very useful, and certainly aid wider communication.[9] He defined algae as organisms that use oxygenic photosynthesis but are not members of the Embryophyta (the group of plants that comprises all the familiar types such as mosses, ferns, conifers and flowering plants). This is effectively the traditional definition of algae, from seaweeds to microscopic green things including cyanobacteria, and is the way I use the term in this book.[10]

Globally, around 50,500 species of algae are known, a figure that includes marine species.[11] 'Algae' is just one of several non-taxonomic terms for groups of organisms I use in this book; others include protists, protozoa and testate amoebae ('testates' for short). None of these groupings makes sense taxonomically, based on our current understanding of evolutionary relationships, but many of them make sense ecologically, grouping together organisms with similar natural histories.

Johann Nave's mid-nineteenth-century guide to collecting algae opened with the claim that 'Few studies afford more instruction, or are attended with more genuine pleasure, than that of the Algae', citing their 'wealth of form' which 'astonishes and delights'.[12] That raises a question – how great is the 'wealth of form' in British non-marine algae? Table 2 attempts to give an overview of the key groups of freshwater and terrestrial algae found in Britain and Ireland. For many groups of organisms that feature in the New Naturalist series, such as birds, butterflies or flowering plants, constructing such a table would be a relatively straightforward undertaking. However, compiling this table was far from simple, and the difficulties are instructive when thinking about our current understanding of algae. Even the high-level groupings – phyla – used are still in flux, and different biologists favour different names for them. In addition, some researchers would lump some

TABLE 2. Key groups (phyla) of freshwater and terrestrial algae found in Britain: based on Canter-Lund & Lund (1995), John *et al.* (2011, 2022), Bellinger & Sigee (2015), Moss (2017).

Phylum	Approximate number of British and Irish species	Typical colour	Visible without microscopy?
Cyanobacteria (blue-green algae) Cyanophyta	360	Blue-green	Some colonial species visible to naked eye (e.g. Figures 79 & 80)
Green algae Chlorophyta	1,110	Bright 'grass' green	Some visible filamentous and colonial species (e.g. Figures 37–39)
Euglenoids Euglenophyta	150	Various colours	All microscopic
Yellow-green algae Xanthophyta	80	Yellow-green	Almost all microscopic, but can occasionally be filamentous or tiny spheres (up to 7 mm)
Dinoflagellates Dinophyta	55	Red-brown	All microscopic
Cryptomonads Cryptophyta	17	Various colours	All microscopic
Golden-brown algae (chrysophytes) Chrysophyta	130	Golden-brown	Almost all microscopic, but can occasionally form visible crusts
Diatoms Bacillariophyta	1,650	Golden-brown	Effectively all microscopic (just occasionally can form visible gelatinous blobs)
Red algae Rhodophyta	24	Red – although many freshwater species are not actually red	Some colonial and visible to naked eye
Brown algae Phaeophyta	2	Brown	Visible to naked eye – multicellular cushions and crustose thalli

Some more obscure smaller groups (e.g. the haptophytes – which have only five British freshwater species) are excluded, and numbers of species in larger groups have been rounded to avoid giving spuriously accurate-looking numbers. This table is not definitive, being very much a summary of a still-changing area of research (see discussion in the main text).

of these phyla together (e.g. golden-brown and yellow-green algae), while others would argue for splitting the large green algal phylum. In the table I have followed the usage in the second edition of *The Freshwater Algal Flora of the British Isles*,[13] as this is the standard key for identifying freshwater algae (except diatoms), and so most likely to be used by any British or Irish naturalist serious about trying to identify algae to species level using microscopy. New species have been added to the list since this was published in 2011 – especially cyanobacteria and green algae.[14] Different authors also disagree about which phyla contain species that can (in mass) form structures visible to the naked eye. I have combined suggestions from several sources, also guided by my own field experience. Therefore Table 2 should certainly be thought of as a work in progress, not a definitive account.

The use of the term 'phylum' in Table 2 and elsewhere in Chapter 2 requires some justification. Algae are classified according to the rules of botanical nomenclature (now formally *The International Code of Nomenclature for Algae, Fungi, and Plants*), and here the level of classification that zoologists call phylum has traditionally been called division. Influenced particularly by Lynn Margulis (who used phylum as a term applicable across all of biology), I have long used phylum instead of division – something that has in the past caused complaints from some more traditional botanists. However, the use of phylum has become more common, and the most important standard source for identifying British algae (excluding diatoms) by microscopy also uses phylum.[15]

One thing that Table 2 does clearly demonstrate is that two groups – diatoms and green algae – are much more species-rich than the other phyla, and so feature repeatedly in this book. Diatoms are described in a bit more detail later in this chapter. Green algae are not just species-rich – with over 1,100 non-marine species in Britain and Ireland, and globally over 11,000 species known to science[16] – but also contain a very large size variation. This ranges from the stoneworts (measured in tens of centimetres) to some of the smallest eukaryotic cells known (although, as mentioned above, it is increasingly common to split this diversity across two phyla, rather than the one used in Table 2). As described in Chapter 3, some of these green algae are only a few micrometres across, overlapping in size with many bacteria, and in some cases small enough for Brownian motion to be an important aspect of their life. Despite their size, very small green algae can be ecologically important, as shown by studies in Llyn Llydaw, only a short distance, as the raven flies, from Llyn Idwal (Figure 40). This nutrient-poor lake has a relatively low density of algae in its waters, and tends to be dominated by tiny green algae – especially the desmid *Ankistrodesmus falcatus* (Figure 41). In other more nutrient-rich lakes in north Wales these small green algae made up a lower percentage of the algal population, with diatoms and cyanobacteria becoming more important.[17]

There is another form in which green algae are familiar to the naturalist, and that is in lichens – which I have discussed in more detail in my previous book in this series.[18] These are composite organisms with a fungus as the main body

FIG 40. Llyn Llydaw photographed from the summit of Yr Wyddfa (Snowdon) in north Wales. The lake is nutrient-poor because the surrounding rocks are hard, and only limited amounts of nutrients are released as they slowly weather. In addition to the small green algae described in the text, *Dinobryon* sp. (Chrysophyta) is an important alga in the spring, with dinoflagellates becoming important during the summer (Priddle & Happey-Wood, 1983).

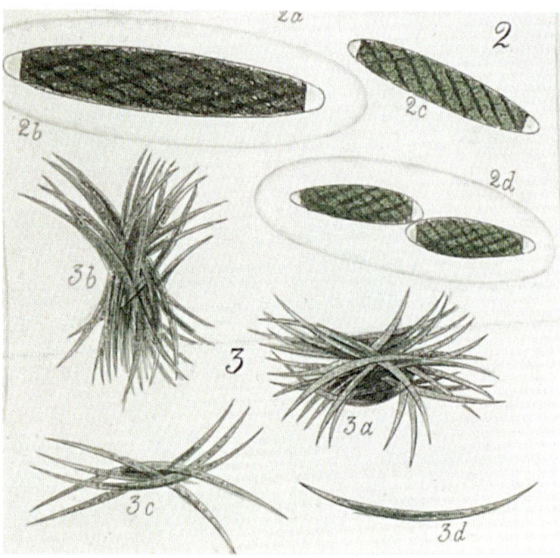

FIG 41. *Ankistrodesmus falcatus* as originally illustrated by Edward Jenner in Ralfs' classic monograph on British desmids (1848). This monograph was the first to use this desmid's currently recognised name. It is labelled '3' on the plate. While the cells of algae in this genus can be between 20 and 165 μm long, they are only 1–5 μm at their widest point (Bellinger & Sigee, 2015). This should make it easy for any nutrients in the water to diffuse into the cell.

of the organism, but also including symbiotic algae which provide products of photosynthesis for the lichen to use (as fungi cannot photosynthesise). In most lichens these photosynthetic microbes are green algae, but in some cases they can be cyanobacterial (as in the lichen *Ephebe lanata*, Figure 36).

The green algae in lichens tend to come from just a few genera, one of the commonest of which is *Trebouxia*. Hidden from sight in the lichen's fungal matrix, there is still much we don't know about these lichen algae. A nice example is provided by the *Trebouxia* algae in the lichen *Ramalina farinacea* – the commonest British lichen in this genus because of its relative tolerance of air pollution. This often contains more than one species of alga, some of which have not been formally described and given scientific names. An example, until recently, was *Trebouxia* species 'TR9', which occurred alongside the formally described *Trebouxia jamesii*. We knew TR9 was widely distributed, as it had been genetically sequenced from multiple sites around the world, but the detailed work needed to describe it formally was only recently done by Eva Barreno and colleagues.[19] They used a mix of microscopic, molecular and physiological data to describe the alga in detail; nicely illustrating the range of techniques that can now be applied to an alga that only a few decades ago would have basically just have been a small green thing seen with conventional light microscopy.

Although *Trebouxia* TR9 is widely distributed around the world, including in several European countries, it's currently not clear if it's found in British lichens. The distribution of the records is mainly driven by the location of the small number of scientists interested in the molecular analysis of lichen algae – rather than being a comprehensive description of where it is found. This is an issue I will return to in the next chapter. However, I have an additional reason for choosing this as an example to describe in some detail – Barreno *et al.* gave the alga the formal name *Trebouxia lynnae* in honour of Lynn Margulis, whose work was described in some detail in Chapters 1 and 2. As they wrote in their paper to explain this choice, Margulis was 'the primary modern proponent of the importance of symbiosis in evolution'. There is a personal link too which drew me to this example. I first got to talk to Lynn at a scientific meeting held in the attractive botanic gardens of the University of Valencia in June 2000, and the lichenologist Eva Barreno was that meeting's local organiser.

Ecologically, the key feature of algae, be they living in the waters of a Welsh lake or in the body of a lichen, is that they usually gain their energy from oxygenic photosynthesis. The same is true of the more familiar green plants too. How algae and larger plants acquired this biochemical trick is a fascinating story that can be illustrated by the testate amoeba *Paulinella chromatophora*.

THE 'MOST INTERESTING' *PAULINELLA*

On Christmas Eve 1894, 25-year-old Robert Lauterborn was collecting samples of diatom mats in an old riverbed just south of Ludwigshafen in Germany. In his samples he found a protozoan that he thought 'in more than one respect can demand to be ranked among the most interesting representatives of its division in freshwater'.[20] When he formally described his find the following year he called it *Paulinella chromatophora*. The generic name *Paulinella* was to honour his aunt and stepmother Pauline (who was one and the same person, as following his mother's death when he was only two his father had remarried, to his late mother's sister).[21] The specific name *chromatophora* was because of two intriguing blue-green coloured structures within the cell.

Lauterborn was right: *Paulinella* does look odd. Most of its cell is filled by these two long blue-green structures. The colour suggests, correctly, that they can carry out photosynthesis; however, the organism is an amoeba and can move about using its pseudopodia. These are the thin structures that can be seen emerging from the shell in Figure 42. However, it seems that it acquires all its energy from photosynthesis, as its pseudopodia appear to be purely used for locomotion, and it has never been observed feeding by phagocytosis.[22] Light can easily get to these photosynthetic structures, as *P. chromatophora* effectively lives in a glass case, with a shell that is made of scales composed of silica (silicon dioxide), which is also the key constituent of glass (Figure 43). When it wants to make a new case, all of the 50 or so new silica scales required are secreted by the protist, and initially stored within the cell. These scales are then manipulated and stuck together to create a new case. The protozoan then divides, and half moves into its new home.[23] Remember – all this complex behaviour, creating an exquisite new case, is being performed by just a single cell.

The natural history of *Paulinella* in the wild is not particularly well known. In general, it seems to live in shady freshwater sediments that may often be low in oxygen, as the microbial decomposition of these organic-rich sediments

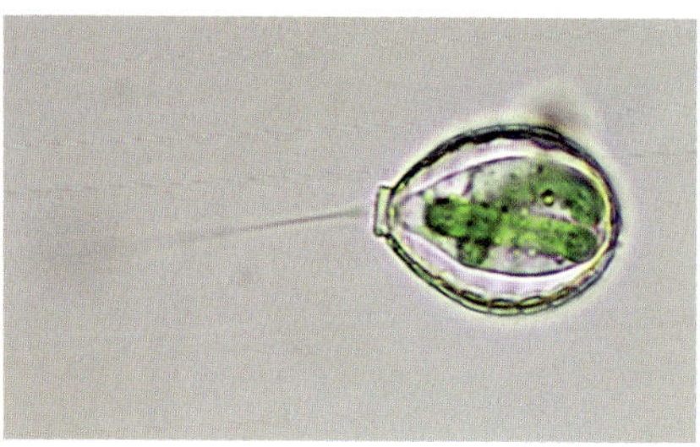

FIG 42. *Paulinella chromatophora* from floating organic-rich sediment collected in spring from a ditch in the Netherlands. Note the two blue-green sausage-shaped photosynthetic bodies, and the thin pseudopodia (filopodia) emerging from the collared mouth of the amoeba's shell. (Ferry Siemensma)

FIG 43. *Paulinella chromatophora* and its habitat at the Lancashire Wildlife Trust's Mere Sands Wood reserve. (a) A rather large empty case from sediments in one of the lakes. (b) Bogbean *Menyanthes trifoliata* at Mere Sands Wood – the sediment surface under these Bogbeans was a *Paulinella* habitat (Creevy, 2013; Wilkinson *et al.*, 2017).

consumes that gas.[24] However, most of what we know of its life in the wild comes from rather anecdotal accounts, as opposed to detailed studies of its natural history. Lauterborn's original habitat of diatom mats matches this idea of it living in organic-rich sediments, as does the site of the example shown in Figure 42. Along with my former Masters student Angela Creevy, I have found it in similar habitats at the Lancashire Wildlife Trust's Mere Sands Wood reserve, where it was in sediments under dense stands of Bogbean *Menyanthes trifoliata* (Figure 43). The suspicion is that photosynthesis is not only supplying energy to the protozoan, but also providing it with additional oxygen, helping it to live in oxygen-poor sites.

However, *P. chromatophora* is most famous for what it tells us about the evolutionary history of photosynthesis in eukaryotic cells. Although our knowledge of its life in the wild is rather limited, there have been more extensive laboratory studies – especially on those intriguing photosynthetic structures, along with studies of how it builds its case. As described in Chapter 2, the chloroplasts found in plants and other photosynthetic organisms are thought to have arisen from the incorporation of cyanobacteria into early eukaryotic cells. All modern chloroplasts can trace their ancestry to this single event, possibly around 1.6 billion years ago, and being a one-off that happened so long ago it's obviously very hard to reconstruct the details. Ideally, we would have a much more recent similar event to study, which might give us more chance of understanding the process.

This is where *P. chromatophora* comes in. The blue-green structures inside it are clearly not chloroplasts. Another possibility is that they are cyanobacteria living symbiotically inside the amoeba. However, they don't look right for that either, as was confirmed by detailed microscopy-based studies in the 1970s. We had to wait until the application of molecular methods in the early years of the twenty-first century to clearly see what was going on.[25] An additional difficulty has been that *P. chromatophora* is really difficult to culture – that is, to persuade it to reproduce in captivity in the lab to provide enough cells for detailed study.

It is now clear that *P. chromatophora* is a very rare thing. It is the only other example we know of where a cyanobacterium has been taken into a eukaryotic cell and has started to turn into a chloroplast-like structure.[26] Remember that the chloroplasts in all other photosynthetic organisms have a single evolutionary origin. In *P. chromatophora*, molecular-clock approaches suggest that the event happened about 90–140 million years ago (well over 1 billion years more recent than the origin of chloroplasts, although a minority of scientists suggest an earlier origin for the *Paulinella* event, over 350 million years ago). In the case of both chloroplasts and mitochondria almost all of the genes that were in the original prokaryote have been lost, with many of them moving to the nucleus of the host cell over the course of evolution. The same has happened with the chromatophores in *Paulinella*, but in this case the process is less advanced and the remaining genome in the chromatophore is still larger than would be typical for a chloroplast. For example, one of the smallest cyanobacteria would have around 2,000 genes, while a green alga would typically have 50–100 in its chloroplast, and for a red alga the figure would be around 200. In contrast, there are still around 900 genes in *Paulinella*'s photosynthetic chromatophores.[27] This starts to give us a rough feel for how long it might take for most of the genes to move to the nucleus – ballpark 100+ million years.

One of the reasons we can be sure that the origin of the chromatophores in *Paulinella* is completely independent of the origin of the chloroplasts in plants is that the genes left in it show that it came from a different group of cyanobacteria. This tells us that different types of cyanobacteria can in principle lead to chloroplast-like structures. It's now known that since the ancestral *Paulinella* acquired chromatophores there has been some speciation, and there are currently two other species of photosynthetic *Paulinella* known (one freshwater, one marine). In addition, non-photosynthetic marine species of *Paulinella*, which feed on bacteria, are also known. Molecular genetic data suggest that the ancestors of these species were never photosynthetic – rather than having lost their chromatophores at some point in their evolutionary history.[28]

The acquisition of a cyanobacterium giving rise to a photosynthetic organelle is referred to as a 'primary endosymbiotic' event. All the members of the Archaeplastida (the kingdom Plantae defined at its most inclusive) are descended from one primary endosymbiotic event that gave rise to chloroplasts and related plastids ('plastid' is the general term for choloroplasts and similar structures in cells). That's what Archaeplastida means – ancient plastids. As described above, *Paulinella* is the only other known example of primary plastid endosymbiosis. However, lots of other types of protists photosynthesise using chloroplasts, a fact that raises the question: how did they acquire chloroplasts?

This is where things start to get rather complicated – and if you are unfamiliar with the various algal groups discussed here, you might like to refer back to Table 2 for a summary of the British groups. Various lines of evidence, both genetic and structural, show that instead of acquiring chloroplasts directly from cyanobacteria, many algae got them indirectly by engulfing other eukaryotic algae that already had chloroplasts. So the cryptomonads and haptophytes acquired their chloroplasts from red algae, while the euglenoids got theirs from green algae. The brown algae and diatoms also had red algae as a source of chloroplasts, while some dinoflagellates got theirs from green algae. These are all examples of secondary endosymbiosis – as there is an additional step between the original cyanobacterium and the final eukaryotic cell.

However, for some groups of dinoflagellates things are even more complex. Some acquired their chloroplasts from the group containing the brown algae and diatoms, which had in turn got their chloroplasts from red algae, whose ancestors had got them from cyanobacteria in the original primary endosymbiotic event. Other dinoflagellates have an equally complex origin for their chloroplasts, initially from haptophytes, which acquired their chloroplasts from red algae too, and so back to the primary endosymbiotic event with the red algae ancestors acquiring engulfing cyanobacteria. To add to the complications, some dinoflagellates are non-photosynthetic and feed on other protists such as diatoms.[29] These diatoms are the most species-rich group of British freshwater algae, and as such are worth describing in more detail, as they feature in multiple examples throughout this book.

DIATOMS: 'NO CELLS MORE BEAUTIFUL'

Diatoms are very species-rich, with around 1,650 freshwater species recorded from Britain (Table 2), and globally at least 30,000 species – with an estimated 100,000 species being considered a more realistic total.[30] They can be found in a wide range of habitats. As Johann Nave wrote in his mid-nineteenth-century handbook on collecting algae, from a time when British amateur naturalists were more interested in the microworld than is the case today:

> *I suppose there is scarcely a single piece of water anywhere which does not contain at least some individuals of the commoner species. They are to be found alike in the lake that crowns the mountain-top, and the swamps and peat-beds which fill the lowest valley; in the watercourse employed to irrigate the meadows; in the broad ocean and the shallow puddle left by the overflowing of a ditch.*[31]

More recent researchers would also point out that they can be found in many soils too.

As with *Paulinella*, diatoms effectively have glass cases made of silica. This is known as a frustule, and as Nicholas Money wrote in 2014, 'There are no cells more beautiful than diatoms in the universe.' This is a big claim for a mycologist, with a side interest in protists, whose field experience is restricted to a single planet. But he has a point – their glass cases are extraordinary, and Money speculates that 'our impression of diatom beauty derives from an innate response to the symmetry of the cells and their elaborate piercings',[32] in the same way that we find the symmetry of many seashells and flowers aesthetically pleasing. The Victorians certainly found them attractive, and you could buy beautiful arrangements of diatom frustules on microscope slides, to view with your home microscope (Figure 44).

This artistic fascination with diatoms is not just a nineteenth-century phenomenon. For example, a diatom photo that was published on the front cover of *British Wildlife* magazine, to illustrate an article of which I was a co-author,[33] attracted the attention of a fabric designer and led to a scarf pattern decorated with stylised diatoms (Figure 45). This book's cover art provides another example. After discussing a couple of possibilities with the artist Robert Greenhalf it quickly became apparent that a design giving prominence to diatoms would give the needed visual impact. A web search for 'diatom art' finds many other examples of both nineteenth-century and more recent artists drawing inspiration from diatoms.

The basic structure of the diatom frustule is a glass-like box formed in two parts. For anyone who has ever spent time in a biology lab, think Petri dish. A diatom frustule similarly goes together with a lid fitting onto a base, with the lid larger than, and so overlapping, the base. This structure has implications for diatom reproduction, as it cannot grow with the organism, a topic I will return to in Chapter 6. Diatoms can be divided into two basic shapes. Centric diatoms (Figure 44) have radial symmetry and tend to look circular when seen from above – very much like a Petri dish. Pennate diatoms (Figure 45) are bilaterally symmetrical,

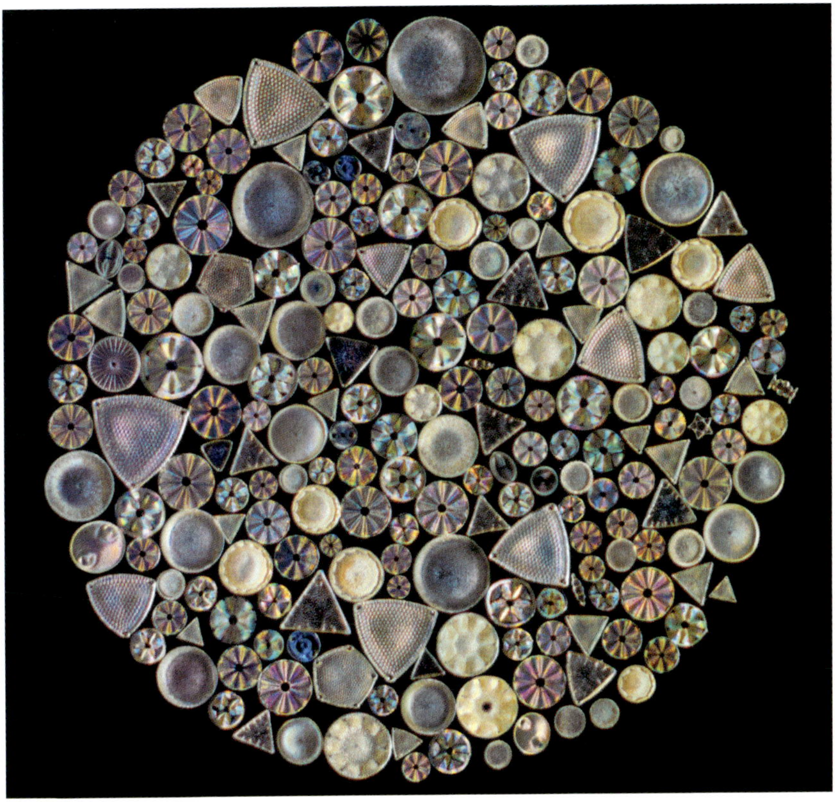

FIG 44. Late-nineteenth-century diatom art. Centric diatoms arranged on a microscope slide, sold by W. Watson & Sons of High Holborn, London. The firm existed from the 1860s until the 1950s, trading as 'W. Watson & Sons' between 1882 and 1908. The diatoms on these slides were usually moved into place using a fine hair (e.g. a cat's whisker), and fixed with egg white or Canada balsam (Money, 2014). (Alex Hyde)

looking long and thin when viewed from above. However, the difference between centric and pennate can be less obvious if the diatom is viewed from the side (often called a 'girdle view'). In some pennate diatoms – including *Pinnularia* (Figure 45) – there is also a raphe; this is a longitudinal canal in the diatom's cell wall which looks like a line running down the diatom when viewed from above with a microscope. This is not present in centric diatoms. In addition the frustule can have a wide range of small markings and pores on its surface, which can be important in identification to species level by microscopy – though they often require treatment of the frustule with acids to make them clearly visible.[34]

FIG 45. (a) A pennate diatom, *Pinnularia viridis*, from the Lancashire Wildlife Trust's Mere Sands Wood nature reserve. This image was used on the cover of *British Wildlife* magazine to illustrate an article on the microbial biodiversity of the reserve (Wilkinson *et al.*, 2017). The magazine was seen by the fabric designer and weaver Janet Phillips, who was inspired to use a stylised diatom motif on a scarf (b), which she also included in a pattern book (Phillips, 2020). (a, Angela Creevy; b, Janet Phillips)

The silica frustule is an unusual feature not seen in most algae, and comes with a mix of advantages and disadvantages.[35] The big advantage is that being mainly inorganic it's much cheaper (i.e. it needs less energy) for the organism to produce, compared with the more organic cell walls of other algae. This can allow diatom numbers in a lake to build much earlier in the year, when conditions are still marginal for photosynthesis. However, as large numbers of diatoms require lots of soluble silica in the water to build their frustules, this can lead to these spring blooms greatly reducing available silica levels in the water column, and so there is often a decline in diatom numbers as the summer progresses. Also, living in a glass box provides another problem for diatoms, making them much denser than water (Box 2). So planktonic diatoms need well-mixed waters, otherwise they just sink to the bottom.

So far, I have described diatoms as what most people would think of as typical algae – photosynthetic microbes that float in the water, or coat the bottom of shallow lakes, streams and other damp habitats. However, they also illustrate that microbial natural history doesn't neatly fit into such straightforward categories. Consider the raphe on some pennate diatoms. In ways we still don't understand in detail, mucus secreted from the raphe allows the diatom to move over solid surfaces, such as rocks in shallow well-lit waters.[37] This seems a much more animal-like, or at least protozoan-like, behaviour. Diatoms also undermine many assumptions about how algae should behave by also being able to use energy from chemicals in their environment – that is, heterotrophy. Again, something that feels more typically 'protozoan' than 'algal'. Over 70 years ago, Joyce Lewin showed that in the lab some species of diatom could grow in the dark using just glucose (not light) as their energy source.[38]

This idea – that many microbes don't fit into the neat binary classification of being *either* photosynthetic *or* heterotrophic ('eating things') – is different from how we think about most larger plants and animals. It will form a recurring theme

Box 2. Density of diatoms and other planktonic algae (mainly based on Dusenbury, 2009).

All planktonic algae (i.e. those that are free-floating in the water) tend to sink, as they are denser than water, unless they have particular adaptations to increase buoyancy. Freshwater has a density of around 1 mg/m^3, while seawater tends to be between 1.02 and 1.03 mg/m^3. A typical algal cell has a density of around 1.03–1.10 mg/m^3 – greater even than seawater – so in very still water most algae will tend to slowly sink. Some algae have evolved adaptations, such as gas vacuoles in some cyanobacteria, to lower their density. Others use low-density fats to help them float. The inorganic silica-rich cases of diatoms give them particular problems, as they can have densities of up to 2.6 mg/m^3, more than double the density of many other algae. Some diatoms will have densities less than this upper limit (down to around 1.4 mg/m^3) as the silicon fraction of the frustule can range from 10–70 per cent silica.[36] These higher densities are why planktonic diatoms need turbulent, well-mixed, water if they are not to sink. Clearly, in deep water, sinking to the bottom may leave them without enough light to photosynthesise.

later in this chapter. But before developing the idea further we need to consider the protozoa – the eukaryotic microbes that are usually associated with movement and heterotrophy, and which therefore used to be viewed as belonging to zoology.

'A WONDERFUL CREATURE IS THE AMOEBA'

Many readers of this book, recalling school science lessons and associated textbooks, may remember being introduced to 'the amoeba' as the key example of 'a simple single-celled organism', or even 'a simple single-celled animal'. In writing this chapter I became interested in how widely the amoeba was used in this way in introductory science classes, so I emailed a variety of colleagues around the world to ask. Clearly it's not just British readers who are likely to remember the amoeba from school science classes – and it's not restricted to English-language countries, or even continental Europe. It's been a key protozoan example in Brazilian school textbooks for many decades, and is also widely used in China – although in China the ciliate protozoan *Paramecium* is also a frequent example in school biology textbooks (*P. aurelia* is probably the second most common protozoan in British school science education).

If that school-book amoeba was identified to species level it will have been called *Amoeba proteus*, and students have been examined on it for well over 100 years. Exactly how long is unclear, as I am not aware of any science historian tracking the early use of *A. proteus* as the archetypal example of a single-celled organism – but in Britain it was certainly widely used in this way by the start of the twentieth century. As recalled by Monica Taylor many decades later, getting examples of amoebae and other organisms for your students to look at was something of a problem, leading to teachers having to 'spend the hours following the dawn tramping the countryside for the collection of biological specimens'.[39] An alternative was a Mr Bolton, a naturalist from Birmingham, who sold tubes of *A. proteus* to educational establishments; but he stopped trading at the start of the First World War, giving science teachers little choice than to go and paddle in ponds to try and find their own. So, around 1900, *A. proteus* was obviously being used in enough science practical classes for it to be worth someone's time to become a commercial supplier. One user of Bolton's amoebae was Lucy Carter at Glasgow University, who, perhaps unsurprisingly, found that there were at least two different species of amoebae mixed together in these commercial amoeba cultures. In fact it was quite hard to successfully culture this protozoan at this time, which is why teachers were having to try and collect it from the wild.[40]

As an aside – it seems that nuns and protozoology went together in Glasgow in the early twentieth century. Lucy Agnes Carter was also known as Sister Bernardine of Jesus, and she was the first woman to be awarded a PhD at Glasgow, in 1921, for her researches in cell biology. Monica Taylor – also mentioned above – was Sister Monica Taylor. She taught science and researched protozoa at a teacher training college in Glasgow through much of the first half of the twentieth century (she was in her mid-eighties when she wrote the paper I cite). In her case it seems that she was able to have a long career in science, despite initial difficulties in getting a formal science education, because of the strong support of Graham Kerr, the influential Professor of Zoology at Glasgow University.[41] Although Kerr specialised in vertebrate biology, he had an 'obsession with microscopy', made great use of protozoa in his teaching, and insisted on teaching women alongside men – rather than in separate single-sex classes. However, there is a story that for a time Sister Monica was forced to attend lectures hidden from the male students behind a curtain until Graham Kerr somehow convinced those in authority to see sense. I know of no contemporary source for this story, but certainly she was required to have a chaperone at lectures, and was only able to attend after Graham Kerr's intervention on her behalf. Although hardly related to protozoology, it's impossible to resist also telling the story that during the Second World War Sister Monica was arrested on suspicion of being a German paratrooper disguised as a nun. She was still attending scientific meetings in her nineties.[42]

One reason that *A. proteus* became so widely used in biology teaching is probably its size. It's just about large enough to be seen as a tiny blob with the naked eye if viewed against a plain background. This means that it's relatively easy for students to see even with a basic school microscope. Size probably also explains why it (or something very similar) was first described as long ago as 1755. There was then over 100 years of confusion about what to call this amoeba, before Joseph Leidy gave the name *Amoeba proteus* to 'the common large amoeba' in 1878. Leidy really liked his amoeba, opening his 1878 paper with the thought that:

> *A wonderful creature is the Amoeba, one of the lowliest of the lowest class of animals, a mere speck of the thinnest jelly endowed with the usual attributes of all living things.*[43]

It's exactly in this sense that *A. proteus* has been introduced to many generations of biology students, except that we would not now call it an animal. At the time, Leidy was one of the most well-known American biologists. He had wide-ranging interests, and although his title was Professor of Anatomy he is often described as a founding figure for vertebrate palaeontology, parasitology and protozoology in North America.[44] The description of the charismatic large fossil mammals and dinosaurs that were turning up in the American West dominated much of the first part of Leidy's career. However, after Othniel Charles Marsh and Edward Drinker Cope entered this field of research, the more retiring Leidy retreated to his earlier interest of microscopy to avoid these 'intensely ambitious men whose behaviour was at best idiosyncratic and at worst simply reprehensible'.[45] Although I have also published on dinosaurs, I have a preference for testate amoebae, and so from my personal perspective the badly behaved Marsh and Cope did protozoan science a great favour by encouraging Leidy to spend more time with his microscope. Among other things, this led to his classic highly illustrated monograph 'Fresh-water rhizopods of North America', which included many beautiful plates – including illustrations of *Amoeba proteus* (Figure 46).[46]

Leidy found *A. proteus* in a variety of 'ponds, lakes and ditches', either on the surface of the sediment or on the underside of the leaves of water plants. The example shown in Figure 47 came from a wet quaking bog in the central Netherlands. In general the habitat seems to be pools or streams, not too stagnant, with reasonable oxygen levels in the water and a supply of amoeba food. Mackinnon and Hawes suggested that good numbers of diatoms and desmids are often associated with good *A. proteus* habitats – both like similar conditions to *A. proteus* and provide it with food.[47] The amoeba also eats other protozoa;

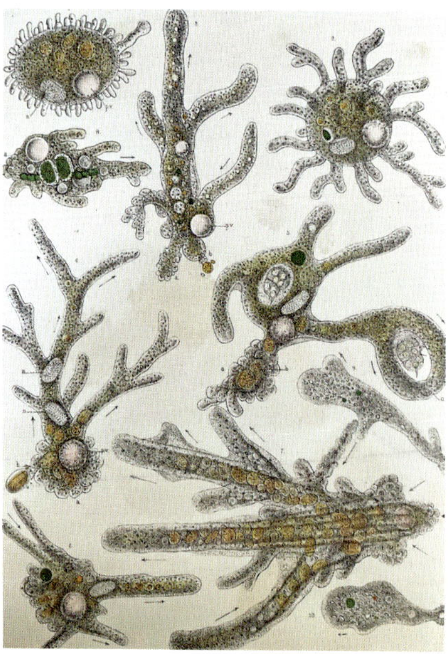

FIG 46. *Amoeba proteus*, from Leidy's 1879 monograph on 'Fresh-water rhizopods of North America' (rhizopod is an old term for amoeba). Leidy considered all the amoebae on this plate to probably be *A. proteus* – except for the two smaller grey protozoa in the lower right part of the plate. Note that the amoeba centre-right is drawn in the process of catching a ciliate in the genus *Urocentrum*; another of these ciliates, already engulfed, can be seen within the body of the amoeba. In the view of Andrew Wier and Lynn Margulis (2000), Leidy's 'drawings of amoebae (rhizopods) are still without parallel'. Given that photomicroscopy has replaced drawing as the main means of recording images of protists, he is perhaps unlikely ever to be surpassed as an illustrator of amoebae.

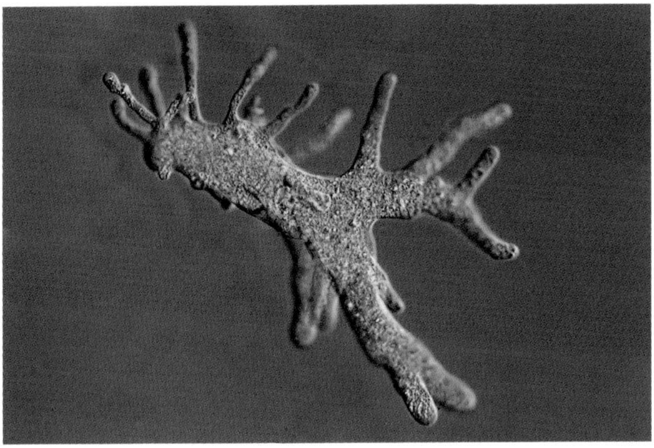

FIG 47. *Amoeba proteus* from a quaking bog at Bert Bospad, a nature reserve in the central Netherlands. This individual was 400 μm long, but this species of amoeba can sometimes grow to around 750 μm (0.75 mm) – a similar size to the full stop (period) at the end of this sentence. The various projections from the main body of the cell are pseudopodia. There are several different types of pseudopodia seen in amoebae – these relatively broad ones are technically referred to as lobose (compare with the thinner filose pseudopodia of *Paulinella* seen in Figure 42). The photomicrograph is by Ferry Siemensma, who took a number of the protozoan photomicrographs used in this book. As a fourteen-year-old he read about *Amoeba proteus* in a popular science book and got a cheap toy microscope to look for it. The toy microscope was not up to the job, but the interest developed (with better microscopes), leading to a long-term interest in protozoa, which he maintained alongside a school teaching career, and has continued into retirement. (Ferry Siemensma)

for example, when Monica Taylor developed a successful culturing operation in Glasgow in the early twentieth century the main food she used was the ciliate *Tetrahymena*.[48] In culture, at temperatures of around 23 °C, *A. proteus* can divide once a day.[49] See the Appendix if you want a simple recipe for culturing amoebae and similar microbes.

THE NATURE OF PROTOZOA

Amoeba proteus is the archetypal textbook protozoan, but in fact there is a colossal, and still poorly studied, diversity of protozoa – that is, protists that feed by phagocytosis (a process known as phagotrophy). In the past it was tempting to describe photosynthetic protists as 'algae' and heterotrophic ones as 'protozoa'. This is still a reasonable starting point, but we have already seen that real-life microbes are more complex than this. *Paulinella chromatophora* is a testate amoeba that has close relatives that seem typically protozoan – that is, they are heterotrophic and feed by phagotrophy; but *P. chromatophora* itself does not use phagotrophy and gets its energy from its photosynthetic plastids. In this section I provide a very brief introduction to protozoan diversity, before describing testate amoebae in more detail in the next section of the chapter.

In the early 1960s – when I was a baby – the textbook higher-level classification of protozoa looked reasonably straightforward. There were four main groups: the naked and testate amoebae (class Rhizopoda), the flagellates (class Mastigophora), the ciliates (class Ciliata), along with the Sporozoa (a class of parasitic amoeba-like organisms).[50] Twenty years later when I was an undergraduate this was still effectively the approach taken by general biology textbooks, although there were differing views as to whether these groups should be considered classes or phyla.[51] However, in the more technical literature it was already clear that these classical protozoan groupings were at best justified as of 'managerial convenience [rather] than anything else'.[52]

We now know that out of these traditional four protozoan groups only the ciliates are monophyletic – that is, they form a group that have all descended from the same common ancestor. The other groupings do not mirror evolutionary history. Flagellates, for example, are spread across at least four supergroups in modern classifications such as that shown in Figure 10.[53] Jump forward another decade, to the time when I was submitting my PhD thesis (on urban trees, not protists), and things had become much more complex. In an article in 1992 Keith Vickerman lamented the decline in the teaching of protozoa in British universities, pointing out that a few years earlier:

> *It was possible to tackle protist diversity, even at first year university level when there were basically four protozoan Phyla, but even in those universities which still afford the protists a home, how can the most motivated students be expected to cope with 45 phyla?*[54]

And this was *before* most of the changes brought about by the extensive modern data on protist molecular genetics! This level of complexity not only makes it hard to teach this area to undergraduates, but is also a considerable problem when it comes to writing a book such as this one. Indeed, I have not even attempted to compile a table for British protozoa to match Table 2 for algae. The necessary information is just not available.

This book's main focus is ecological rather than taxonomic, and there are two of these traditional protozoan groups that particularly feature in the examples used – ciliates and amoebae. The ciliates make sense taxonomically, as they form a monophyletic group. Ciliates (most commonly now placed in the phylum Ciliophora) are single-celled heterotrophic protists, and get their name from short hair-like structures called cilia, which are used in locomotion and for feeding. These cilia are similar to but shorter than the flagella of protists such as dinoflagellates. A typical ciliate will have a dense covering of cilia, often arranged in rows. The commonest food for ciliates is bacteria, but some eat other protists, or feed on dissolved nutrients in nutrient-rich waters (osmotrophy), and they can be found in both water and soils;[55] indeed, ciliates from very arid African habitats were described in Chapter 2. Some ciliates contain photosynthetic algae (Figure 48), which allows them to function simultaneously as both algae and protozoa – more formally, it allows them to be both autotrophic and heterotrophic. Ciliates have also long been known to indulge in sexual reproduction, while many other types of protozoa have traditionally been assumed to be asexual. This is a topic I will return to in Chapter 6.

The ciliates have also played a role in the development of ecological science, being used in several classic experiments because they are small predators that reproduce rapidly. This means that you can run in a few weeks on a lab bench an experiment that would take years, and much more space, if you used larger organisms. For example, in the 1960s Vandermeer used small test tubes with

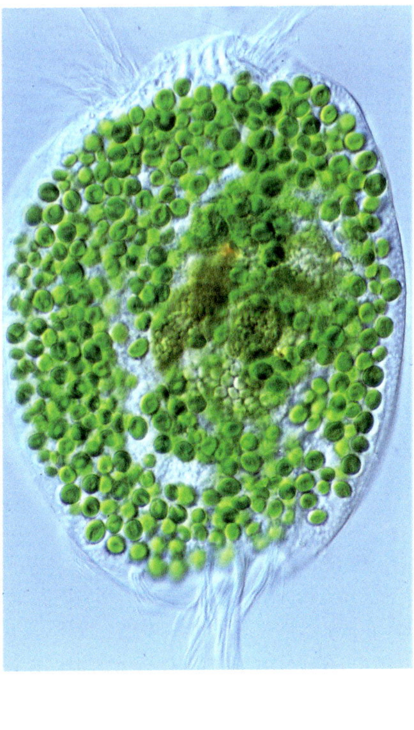

FIG 48. *Euplotes daidaleos* a ciliate protozoan with endosymbiotic green algae (*Chlorella*) from Priest Pot in the English Lake District (see Chapter 5 for more details of this site). The algae live inside the ciliate, effectively making the ciliate photosynthetic. (Genoveva Esteban)

experiments running for around a month – think of the much larger size of experiment you would need if using even small animals.[56] Various species of *Paramecium* have often been the organism of choice in such experiments, with competition between the protozoa being a key idea under investigation.[57] The work by Georgii Gause in the 1930s is still often described in modern textbooks on ecology, and I have written about it in more detail in a previous book in the New Naturalist series.[58]

The amoebae, unlike the ciliates, no longer hold together as a taxonomic group. The key feature of amoeboid protists is that they are unicellular eukaryotes that use pseudopodia for feeding and locomotion. These pseudopods are temporary extensions of the protozoan's cell, and can clearly be seen in the photomicrograph of *A. proteus* shown in Figure 47. Such amoebae are mainly phagotrophic, and can be very common in soils and water – where they can use their pseudopodia for anchoring themselves to sediments. There are also parasitic species (such as those that cause malaria), but these fall outside the remit of this book.

It is now known that amoebae are scattered widely across the eukaryotic tree of life, being found in multiple supergroups (Figure 10) – especially the Amoebozoa and the Rhizaria.[59] Traditionally the amoebae were split into naked amoebae and testate amoebae (with 'shells' usually referred to as tests). The textbook icon *Amoeba proteus* is a naked amoeba, and somewhat simpler than most of the Amoebozoa. As well as tests, some amoebae – the amoeboid flagellates – have flagella. Others can form cysts that can survive harsh conditions in a dormant state. Yet others have a stage in their life cycle where the cells gather together as a multicellular organism, which can produce fungal-like fruiting bodies that release spores (these are the slime moulds discussed in more detail in Chapter 7). *Amoeba proteus* has no test, and is not known to form cysts or spores, or to have a multicellular stage in its life cycle.[60] As with the frustules of diatoms, the tests of testate amoebae make them particularly amenable to study by simple microscopy, and they are worth describing in a bit more detail.

TESTATE AMOEBAE

Testate amoebae are found in at least three major groups of protists. Even before modern molecular data were available it was obvious that rather different types of protozoa were included within the 'testate amoebae' grouping, because different species had different types of pseudopodia (Figure 49). So rather than being defined by a shared evolutionary history, testate amoebae are defined by their

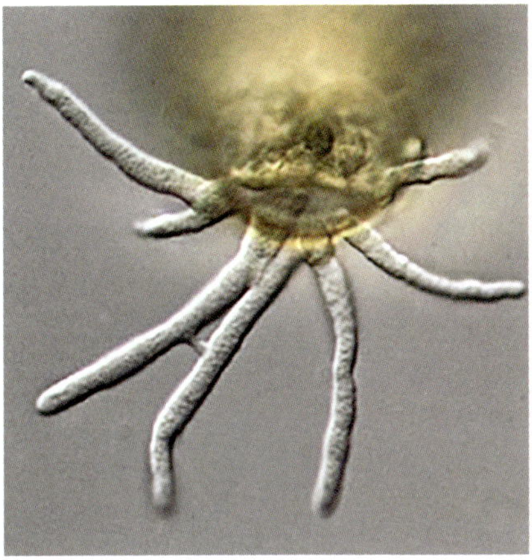

FIG 49. Lobose pseudopodia of the testate amoeba *Heleopera sylvatica* (an amoebozoan testate amoeba). Compare this with the filose pseudopodia of *Paulinella chromatophora* shown in Figure 42 (this is in the SAR group – see Figure 10). Even before modern molecular evidence was available, these differences in pseudopodia showed that there were at least two very different groups of testate amoebae (Ogden & Hedley, 1980); however, only with molecular data did it become apparent just how distantly related these groups were. (Ferry Siemensma)

test – although by convention not all shelled protists are included as testates (for example the Foraminifera – an important marine group – are excluded). The tests can be composed of organic matter such as protein, glass-like scales made of silica or, alternatively, they can be constructed out of mineral grains and other fragments taken from the environment – in similar fashion to the cases of many caddisfly (Trichoptera) larvae.

Most testates are from freshwater or terrestrial habitats, bogs with *Sphagnum* mosses being a particularly rich and well-studied testate habitat (Figure 50).

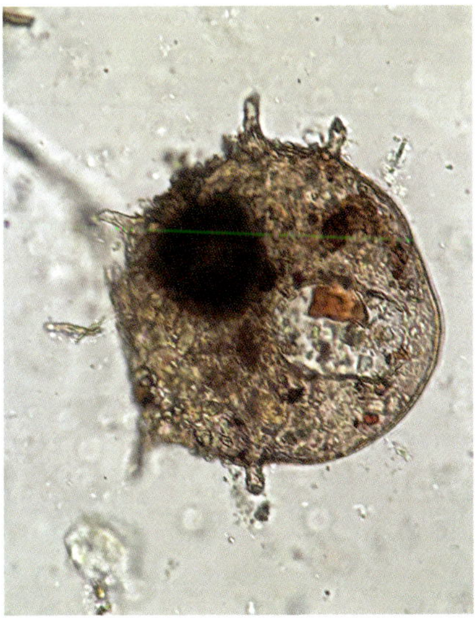

FIG 50. The testate amoeba *Centropyxis aculeata* (typically 70–180 µm across), a species found associated with bog-moss *Sphagnum* and other wet sites. The test is made of protein but often covered, as here, with lots of small mineral particles. The pseudostome (the 'mouth' through which the pseudopodia emerge) can be seen as a lighter-coloured aperture to the centre right. The number and arrangement of the spines on the testate can be rather variable (Chardez, 1970). These spines may help anchor the test in very wet conditions with flowing water. This example came from flat-topped bog-moss *Sphagnum fallax* on the edge of a small tarn in Ennerdale in the English Lake District.

However, there are interesting, and seldom studied, testates from marine beach sand too, although as salinity increases the number of testates tends to decrease. Until the recent work by Buckman and colleagues, the only study of beach-sand testates in Britain I am aware of was a short paper in French published some 40 years ago.[61] Indeed, away from peat bogs the number of British testate studies is quite limited.

The tests of amoebae in the genus *Difflugia* are of particular interest, as these are the ones that most call to mind the construction of caddis cases. The complexity of these cases, built by a single-celled organism, demonstrates that extremely complex-looking behaviour doesn't necessarily need complex intelligence. For this reason, *Difflugia* tests attracted the interest of Mike Hansell, who, after doing a PhD on caddis case construction in the 1960s in Niko Tinbergen's group at Oxford, became an expert on animal architecture at the University of Glasgow. In his book *Built By Animals*, written in retirement, he opened a chapter entitled 'You don't need brains to be a builder' with a discussion of *Difflugia* tests.[62] Early applications of electron microscopy to *Difflugia* tests had shown that amoebae seemed to be rather selective in their choice of materials used, and had suggested that 'shell construction must be a complex activity'.[63] Subsequent research has only emphasised the complexity of what these amoebae are doing.

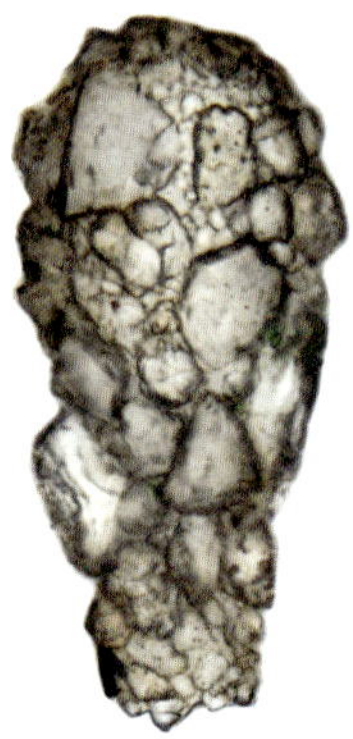

FIG 51. Test of *Difflugia oblonga* (typically 100–300+ µm long). *Difflugia* is a very large genus, with considerable uncertainty associated with the status of many of the, often very variable, species. Only very recently have molecular data started to be used to refine the classification of these amoebae (e.g. González-Miguéns *et al.*, 2022b), and *Difflugia* taxonomy is likely to continue to change in the future. (Ferry Siemensma)

Difflugia oblonga (Figure 51) has a rather variable test, which creates some problems with identification. It is associated with wet sites – from freshwater to bogs and wet soils – and its test building was the subject of an interesting 2013 study by Eric Armynot du Châtelet and colleagues.[64] Using state-of-the-art microscopy, including x-ray microtomography – which effectively takes a large number of microscopic x-rays and then stacks all these together to give a 3D image – the researchers showed that this testate was not just using the commonest mineral grains in its environment to construct its test. Instead, it was selectively finding smaller particles, especially low-density silicates, for use in test construction. The maximum size used seemed to be governed by the size of the test's mouth, as the particles were taken into the test before being rearranged to construct a new shell. *Paulinella*, described above, does something similar, but with internally secreted scales rather than material collected from its environment. Remember, as Mike Hansell stressed, all this is being done by an organism without a brain – indeed, by a single cell.

An obvious question is: what are these various tests for? What is the adaptive value of their shells to testate amoebae? One suggestion is that they are for protection, either as a defence against predators or as a shelter against environmental stress such as drought. In Chapter 2 I have already pointed out that the first Precambrian testate-like organisms (the vase-shaped microfossils) sometimes have holes in them suggesting predator attack. As these were clearly aquatic organisms, probably marine, drying out is unlikely to have been an issue for them – so predation may have been a key factor in the evolution of their tests. Protecting the amoebae against parasites is also plausible. By the time we see fossil testate amoebae on land, as in the Rhynie chert, viewing the tests as a defence against drying out seems plausible.

However, there are other possible explanations too. In the early 2020s, Money and Fischer suggested that by increasing the density of the amoeba, the test would help it maintain contact with surfaces rather than being dislodged by turbulence in the surrounding water (remember that many testates in soils and leaf litter are often operating in a film of water that coats the particles to which they are attached).[65] Money and Fischer presented back-of-the-envelope calculations that suggest a testate may have a density 50 per cent greater than a similar-sized naked amoeba. Recently, another suggestion has been made, for the arcellinid testate amoebae (that is, those in the Amoebozoa). Kenneth Dumack and colleagues have plausibly proposed that the shell helps these amoebae attack larger prey, for example providing points of leverage that aid in breaking open diatom frustules.[66] It is probably a mistake to ask which of the explanations is the correct one. Several of these possibilities may have contributed to the evolution of amoeba tests, and it's also plausible that the initial reason for their evolution could be different from the main advantages they give today.

As mainly phagotrophic protozoa, the question of what testate amoebae eat is clearly of some importance. Much of what is eaten is, unsurprisingly, rather small – such as bacteria, small algae and other protozoa.[67] But small animals of the microworld can be eaten too. For example, planktonic *Difflugia tubersinifera* collected from Chinese reservoirs were recorded eating rotifers.[68] I have also seen rotifer trophi (their 'jaws' – the only hard part of a rotifer) in the cell

OPPOSITE: **FIG 52.** *Hyalosphenia papilio* illustrated by Joseph Leidy (1879). The green spheres are endosymbiotic algae living within the protist. Molecular data have shown that *H. papilio* actually comprises multiple closely related types, which are effectively impossible to tell apart using microscopy (Heger *et al.*, 2013). This species was one of the first amoebae that Leidy found, and became one of his favourites. He fondly remembered the field sites among the bogs and swamps of New Jersey where he first found it, 'localities [which] have special charms for the botanical student on account of the diversity of beautiful and interesting plants'. And he describes how, 'during the Christmas holidays, I have repeatedly exhibited it, in the living condition, to the admiration of friends' (Leidy, 1879, p. 132).

contents of *Nebela tincta* from terrestrial moss samples, although in this case it was impossible to know if it had captured a live rotifer or scavenged a dead one.[69] Nematodes are also eaten by larger testates,[70] and there are even records of groups of small testates – *Cryptodifflugia operculata* – overwhelming much larger nematode prey. I will return to testate feeding, including this apparent pack hunting behaviour, in the context of nutrient cycling in Chapter 8.

Many testates can also photosynthesise by proxy, using endosymbiotic algae. A good example is *Hyalosphenia papilio,* a common testate in *Sphagnum* bogs (Figure 52), which is associated with green algae in the genus *Chlorella.* Such algal symbionts can be seen in many other species of testate amoebae. Interestingly, sampling of such algae from very distantly related testates showed that the algae were all closely related to each other, despite the protozoans being from different eukaryotic supergroups. These have been informally named TACS (testate amoebae *Chlorella* symbionts).[71] The fact that they are closely related suggests that these algae are in some way adapted to the symbiotic way of life. Indeed, they are similar to *Chlorella,* which are also endosymbiotic in some species of *Paramecium,* and they are also similar to some lichen-associated algae.

MULTIPLE WAYS OF MAKING A LIVING

In the macroworld we are used to organisms that usually make their living either by photosynthesis or by heterotrophy (eating other organisms) – e.g. cabbages or caterpillars. Carnivorous plants, which do both of these things, tend to strike us as rather unusual. Similarly, microbial protists are traditionally described as algae (photosynthetic) or protozoa (heterotrophic), but in the microworld things are often much more blurred. Mixotrophy (using both photosynthesis and heterotrophy) is much commoner at this smaller scale, and multiple examples from diatoms, ciliates and testate amoebae have been described in this chapter.

Even if you choose to view the definition of plants and animals as ecological, rather than taxonomic, the microworld is full of species that just don't fit this longstanding binary division. Some of these organisms have in the past been treated as both algae (by botanists) and animals (by zoologists), a good example being the euglenoids (see Table 2), which botanists have placed in the phylum Euglenophyta, while zoologists have classified them as Euglenozoa.

I have picked one group of algae (diatoms) and one group of protozoa (testate amoebae, a non-taxonomic grouping) to describe in more detail, as these are particularly amenable to study with very limited lab facilities. They can be extracted from field samples and viewed with a basic compound microscope, without the need for culturing or other more complex techniques (see Appendix). In addition, they can be very beautiful when viewed with a microscope. Also, because they don't need culturing, they can be counted on a microscope slide in a similar way to how you might count ducks on a lake, or trees in a hedgerow (see Figure 11). This makes them interesting groups for ecological research. In addition, their shells can survive entombed in lake sediments and peats, allowing their remains to be used to investigate past environmental changes. These multiple advantages explain why diatoms and testates feature in many examples throughout this book.

The example of *Amoeba proteus,* and its long history of use in biology teaching, also illustrates the idea of using protists as model organisms. That is, organisms that seem particularly appropriate for studying some aspect of biology – such as the Fruit Fly *Drosophila melanogaster* widely used in genetics experiments. Away from *A. proteus*, examples in this chapter include *Paramecium* in ecology experiments, and *Paulinella chromatophora* in studying the evolution of the chloroplast. Not only have algae and protozoa provided several such model systems in the past, but there is lots of scope for them providing a wider range of model organisms in the future.[72] With most of biodiversity being microscopic, there is a great danger of being led astray if we only rely on large model organisms, such as *Drosophila* or Thale Cress *Arabidopsis thaliana* (if you think of these two species as small, then you have a very biased view of biodiversity). This chapter has provided an introduction to the main types of algae and protozoa. In the next, I will consider wider aspects of their diversity: their numbers, distribution, dispersal and species-richness.

CHAPTER 5

Diversity and Distribution: the View from Rochers-de-Naye

It's mid-morning in late June and, having found the right platform at Montreux station, I have just boarded the train for the summit of Rochers-de-Naye. I am having a day off botanising during a sabbatical visit to the École Polytechnique Fédérale de Lausanne to work on testate amoebae. Over the next hour the train takes me slowly up the scenic rack-and-pinion railway to the top station, just below the summit, which gives a panoramic view over the Swiss and French Alps. To a Brit it's all a bit reminiscent of the Snowdon Mountain Railway, and as on Yr Wyddfa (Snowdon) the train feels a bit like cheating, but it's the most practical car-free way to get up high to an abundance of alpine flowers.

FIG 53. Alpine Pasque Flower *Pulsatilla alpina* below the summit of Rochers-de-Naye, Switzerland.

This far into the year almost all the snow has gone and the flowers have taken over. Although almost twice the height of Yr Wyddfa, Rochers-de-Naye shares the good-weather crowds of people around the top station and the summit. But, just as on Yr Wyddfa, away from the top there are fewer people but better plants, and the place starts to feel more like a mountain and less like a tourist destination. My timing is perfect for the Alpine Pasque Flowers *Pulsatilla alpina* (Figure 53), and indeed the whole collection of flowering alpines is more showy and spectacular than I am used to from the British mountains.

As well as the Alps and the alpines it's the view of Lac Léman that dominates the landscape, sunshine-blue on this fine June day (Figure 54). The lake is huge by European standards, fading into the distant haze towards Geneva. It also has an important place in the history of research on the diversity of amoebae.

FIG 54. Lac Léman (Lake Geneva), the largest lake in Switzerland, photographed from close to the summit of Rochers-de-Naye. The amoebae of this lake, and other wet habitats in its catchment, were described over 100 years ago in a major monograph by Eugène Penard (1902).

OPPOSITE: **FIG 55.** Figure from Penard (1902) illustrating all 17 known species of testate amoebae that he had not found in his studies of Lac Léman. Number 14 is *Paulinella chromatophora* (described in Chapter 4), which he presumably missed through not sampling a suitable habitat for it (although he did sample some small pools in the lake's catchment that might have been suitable). One habitat he did not sample was sandy beaches. When Vassil Golemansky and Milcho Todorov (2009) did examine this habitat at Lac Léman, around 100 years later, they found 18 species of testate amoebae – two of which were specialists usually only found in this habitat, and so not found by Penard. Several other species were not in Penard's list, including some not known to science at the time of Penard's studies (Penard himself described over 30 new species of amoebae during the course of his Lac Léman work). *Nebela ansata* – now called *Cornutheca ansata* (number 4 on Penard's figure) – is of particular interest as having a very limited distribution on the east coast of North America, and is described in more detail later in this chapter.

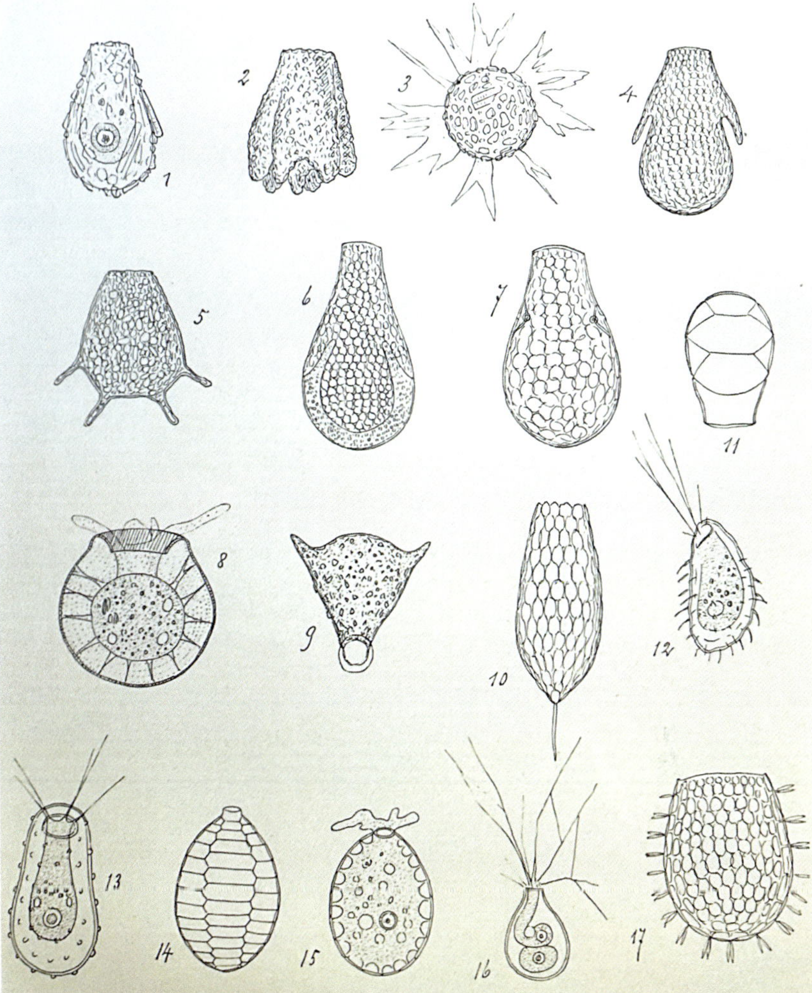

Rhizopodes absents jusqu'ici du territoire exploré. — 1. *Difflugia asterisca.* — 2. *Difflugia fragosa.* — 3. *Difflugia acropodia.* — 4. *Nebela ansata.* — 5. *Nebela caudata.* — 6. *Nebela hippocrepis.* — 7. *Nebela bigibbosa.* — 8. *Arcella mitrata.* — 9. *Campascus cornutus.* — 10. *Euglypha mucronata.* — 11. *Sphenoderia macrolepis.* — 12. *Pamphagus armatus? (Trinema spinosum).* — 13. *Trinema verrucosum.* — 14. *Paulinella chromatophora.* — 15. *Pamphagus Dittrichii.* — 16. *Microgromia socialis.* — 17. *Placocysta spinosa.*

In 1902 – just a decade after the rack-and-pinion railway arrived at the summit – Eugène Penard published a large monograph of 714 pages on the amoebae he had found in and around the lake.[1] Penard was a native of Geneva and, like Joseph Leidy, he is a major historical figure in the more descriptive aspects of protozoan natural history. However, unlike Leidy, he never managed to secure an academic position.[2]

One of Penard's conclusions in his Lac Léman monograph was very different to what we are used to with larger organisms. He wrote that 'after two years of continuous methodical work I found in the Lake Léman basin approximately 92 per cent of the species described in the world'[3] – although he goes on to point out that this figure applies only to what we would now call testate amoebae, so excluding the naked amoebae. To emphasise the point, he includes a figure that illustrates *all* the then known species of testates *not* found in his Lac Léman study (Figure 55). Contrast this with the mountain plants I was looking at. If you travel the world to mountains with similar environments you expect to see many different plant species – that's why you would take different field guides if you were visiting the Himalayas or the Andes, rather than the Alps. This is true for all the groups of larger organisms we are familiar with. Was Penard's 92 per cent figure wrong – an artefact of our limited knowledge of amoebae at the start of the twentieth century? Or is there something very different about global diversity in the microworld? This is a key question in the natural history of microbes, and one we still do not have a definitive answer to. A good place to start considering these, and related, questions in more detail is the English Lake District.

THE VIEW FROM PRIEST POT

At the northern end of Esthwaite Water, to the west of Windermere, is an expanse of fen vegetation known as North Fen. This was a classic site of early British plant ecology, studied by W. H. Pearsall around the time of the First World War and later described in detail in Tansley's monumental synthesis of British vegetation.[4] This fen has infilled the northern end of Esthwaite, separating Priest Pot from the rest of the lake, a separation that happened at least 400 years ago. The Pot currently has a surface area of around 1 hectare, and a maximum depth of just under 4 metres. Today it is surrounded by woodland, to the extent that in summer it's practically invisible from the road. This has important effects on its ecology, as the trees shelter it from the wind. The lack of wind means that there is relatively little mixing of the water, and so in summer thermal layers tend to form – with warmer water sitting on top of cooler water. This in turn leads to the bottom waters being very low in oxygen, something normally associated with much deeper lakes. Photographs from 1900 show almost no trees surrounding the lake, so this aspect of its ecology is likely something that has developed over the last 100 years or so. The Pot is also relatively nutrient-rich – often described as a 'productive pond', unlike many of the other Lake District lakes and tarns, which are typically nutrient-poor.[5]

Priest Pot is an important site for anyone interested in the natural history of freshwater protists in Britain, as from the mid-1980s right through the 1990s it was intensively studied by Bland Finlay and his collaborators (Figure 56). At the time

FIG 56. Sampling Priest Pot in the late twentieth century. (a) Bland Finlay and his technician Ruth Hindle using winter conditions to sample Priest Pot through holes in the ice. Probably taken early in 1986 – February that year was exceptionally cold. (b) Visiting German Professor Walter Traunspurger and colleague sampling nematodes at Priest Pot. (a, U.-G. Berninger; b, Bland J. Finlay; both photos kindly provided by Genoveva Esteban)

they were based at the Institute of Freshwater Ecology, located in the Freshwater Biological Association (FBA) laboratories, Windermere. A very useful summary of all this work can be found in a booklet by Finlay and Maberly, published by the FBA in 2000.[6] These studies of protist diversity at Priest Pot relied on microscopy, as it is only more recently that suitable molecular methods have developed for use in this type of work – that is, identifying the microbes by their genetic material in water samples rather than visually identifying them under a microscope.

By attempting to identify as many as possible of the microorganisms living in this Lake District pond, Bland Finlay and colleagues were able to attempt a modern version of the global comparison Penard carried out for Lac Léman at the start of the twentieth century. This work also allows us to attempt an answer to another basic question: how many microbial species live in such a pond? Finlay and Maberly gave an 'interim list' of all of the microbes then identified from Priest Pot – 'interim' as they were sure the list was incomplete, despite this being an unusually well-studied site.[7] The list included 94 species of bacteria, 16 microfungi and 611 protists. The largest group of protists was the ciliates (256 species), followed by the golden-green algae (96), diatoms (83) and green algae (54). However, they pointed out that these protist groups were also the ones that they had studied in most detail. There were also 57 species of amoebae recorded (34 naked amoebae and 23 testates).

This work makes Priest Pot one of the best-studied sites for freshwater protists in Britain, if not the world. Because of this it's recently been sampled by a group from Oxford University led by Tom Richards, searching for freshwater protists for the Darwin Tree of Life project with the aim of sequencing the full complement of the cell's DNA (its genome) from as much of this diversity as possible. This would have been impossible in the 1990s when Bland and colleagues were working at the Pot. At that time just sequencing the genome of a single species – ourselves – required a huge decade-long international effort. The speed with which this technology has developed is extraordinary. Analysis of the Priest Pot genome samples is currently ongoing.

The Priest Pot results from the 1990s start to give a feel for the sort of diversity one might expect in a freshwater habitat, but what about soils? Some years later the same team attempted to answer this question with an intensive three-year study of soil protozoa (so no algae) from a similar-sized hectare of upland grassland at Sourhope in southern Scotland.[8] Again using a microscopy-based approach, they found 365 protozoan species. Using the traditional major protozoan groupings (though only ciliates are monophyletic – see Chapter 4), this gave 110 species of testate amoebae, 102 ciliates, 100 flagellates and 53 naked amoebae. Some of the species were found both in Priest Pot and in Sourhope soils – such as the testate amoeba *Euglypha rotunda* (Figure 57). Ken Clarke, the

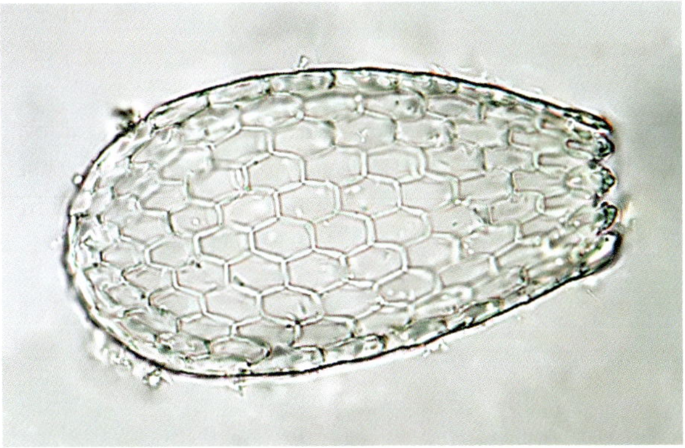

FIG 57. The testate amoeba *Euglypha rotunda* was one of the commonest testate amoebae in the Sourhope soils, and also found in Priest Pot. This morphospecies is very widely distributed – indeed, I am surprised when I don't see it in a sample, almost irrespective of the habitat or continent that the sample came from. However, it has been clear for some time that *E. rotunda* is likely made up of several different species that appear similar, or even identical, under light microscopy. For now, ecological studies are still using this name despite the knowledge that it's not a 'good' species, making it a so-called shadow species – one that is known to need taxonomic revision but the necessary work has not yet been done (Soler-Zamora *et al.*, 2023). (Ferry Siemensma)

group's electron microscopist, took a particular interest in the testate amoebae, and produced a useful illustrated guide to soil testates based on the species they found at Sourhope.[9] One obvious difference between the two sites was that the amoebae, especially the testate amoebae, were much more diverse in the soil study compared to the dominance of ciliates at Priest Pot. There are several explanations for this, one of which is that while the aquatic ciliates had been studied in particular detail at the Pot, the habitats that were likely to be most rich in testates (such as lakeside mosses) were not sampled.

The ciliates in Priest Pot were studied in such detail in order to compare the diversity in this small 1 ha pond with global freshwater ciliate diversity, effectively a modern attempt to revisit Penard's 92 per cent conclusion (but for ciliates instead of testate amoebae). In January 1996 Tom Fenchel and colleagues used a boat to take a sediment sample from the deepest part of the Pot (3.5 m water depth), and analysed it back in the lab.[10] Initial examination revealed only 20 species of ciliate protozoa. However, any that were present as resting cysts would not have been found, and additionally, if the species was really rare in the sample the likelihood of finding it would also be low.

In an attempt to compile something closer to a full list for all the species in the sample, an enrichment-culture approach was adopted. The idea of enrichment cultures is that you can potentially find different species of microbes by making the conditions more to their liking. This was an idea developed by early microbial ecologists such as Sergei Winogradsky and Martinus Beijerinck at the end of the nineteenth century.[11] Tom Fenchel and his team incubated subsamples of sediments under a variety of different conditions (such as different temperatures, pH, food sources, light levels) over a period of 130 days.[12] By the end of this they had identified an additional 115 species that they hadn't found in their initial observations of freshly collected samples. They estimated that this gave them approximately 75 per cent of all known free-living freshwater ciliate species from one small sample taken from the bottom of just one Lake District pond. For comparison, in the Sourhope soils they estimated that they had found around 25 per cent of global free-living protozoan species. These results were to become very influential in debates about the nature of the global diversity of free-living microbes. But to give these a grounding in theory we need to go back to an important book on microbial ecology, published in Dutch in 1934.

EVERYTHING IS EVERYWHERE, BUT THE ENVIRONMENT SELECTS

A longstanding idea in microbial ecology is that microbes are so numerous, and so small and easily dispersed, that any species of microbe can reach anywhere on the planet. However, the microbes you actually find at any one site are a subset of global microbial diversity selected for by the environmental conditions at that site. So while all microbial species can reach a site, only those with suitable adaptations for the local conditions can actually grow, and are available to be found by a naturalist sampling that location. This idea is often described by the phrase 'everything is everywhere, but the environment selects'.

The origin of the idea in this form is a fascinating small book published in 1934 by Lourens Baas Becking – a professor at the University of Leiden, and a key figure in the early history of microbial ecology.[13] Because the book was published in Dutch, there has been some confusion over the years among non-Dutch speakers about what Baas Becking actually said. In his opening chapter he wrote '*Alles is overal*, maar, *het milieu selecteert*' – '*Everything is everywhere*, but, *the environment selects*' – the italics are in the original Dutch version.[14] In a review published in 2006, De Wit and Bouvier argue that the italics (and specifically the lack of italicisation of the Dutch word *maar*, meaning 'but') are important.[15] They point out that what Baas Becking intended to imply by stressing 'but' in this way was that the second half of the phrase appears contradictory to what is implied in the first half. So although all microbial species are at every site, only those that encounter the correct conditions for growth and reproduction will be there in numbers that make it likely you will find them – that is, will be selected for by the environmental conditions.

In his notes added to the English translation of Baas Becking's book, Don Canfield describes 'everything is everywhere ...' as 'one of the most famous quotes in all of microbial ecology'. The idea was not completely new to Baas Becking, but his succinct phrasing became very influential. Some 20 years earlier Martinus Beijerinck in Delft in the Netherlands had made effectively the same point, albeit less succinctly. He was one of the key figures in developing the enrichment culture method, and, as with the work at Priest Pot, found that he could often make a long list of microbes appear that were not initially apparent in his samples.[16] That is, he was experimentally altering the environment to select different species.

The idea that free-living microbes were globally distributed had been around for some time. As already described at the start of this chapter, Eugène Penard thought that most freshwater amoebae were global in their distribution at the start of the twentieth century. A similar conclusion was reached by Harold Sandon, working at Rothamsted Experimental Station near London in the 1920s. In what was probably the first book on the global distribution of soil protozoa he concluded that 'all the soil protozoa appear to be world-wide in their distribution, the same species occurring in arctic, temperate, and tropical soils'.[17] Baas Becking's famous phrase expands this idea by drawing attention to the selection of species by the environment.

By the end of the twentieth century, Bland Finlay became a key proponent of the 'everything is everywhere' idea, in particular writing a very influential essay in the leading international journal *Science* arguing that there was a global dispersal of free-living protists, very different from what we see for larger organisms.[18] This was controversial, and in particular Wilhelm Foissner, another ciliate specialist, was at that time arguing that many protists had limited distributions. Foissner eventually came to the conclusion that in reality there was 'moderate endemicity' – with microbes tending to be more widely distributed than most larger organisms but often falling short of the full idea of 'everything is everywhere'.[19] One important outcome of this dispute was to draw attention to the importance of dispersal in understanding microbial diversity, at both local and global scales.

DISPERSAL: THE VIEW IN INGLEBOROUGH CAVE

Ingleborough Cave is a short walk from the Yorkshire village of Clapham, following Clapham Beck along a pleasant wooded dale. The valley's botany is interesting, and sometimes surprising, as a former owner of the land was a well-travelled Edwardian plant collector who brought many species back from his travels. The cave itself is one of three famous tourist show caves in the limestone of the Yorkshire Dales National Park. All of these show caves have examples of lamp flora (often referred to by the German term *Lampenflora*), growing near the lights that illuminate the cave for visitors. In Ingleborough Cave these lamp floras are mainly microbial biofilms, especially comprising cyanobacteria and diatoms – adding various shades of green to the wet cave walls (Figure 58). In Stump Cross Cavern and White Scar Cave, as well as these microbial biofilms there is also a greater range of larger plants such as mosses, and even the occasional tiny fern, growing around the lights often hundreds of metres from

FIG 58. Lamp flora in Ingleborough Cave, in the Yorkshire Dales. Algae, probably mainly cyanobacteria, growing on flow stone next to one of the lights in the cave roof that illuminate this show cave for visitors. In a study of the Ingleborough Cave lamp flora Allan Pentecost (2010) recorded four species of cyanobacteria, two species of diatoms, some unidentified green algae and unidentified bryophyte protonema (thread-like structures produced by germinating spores).

the entrance. In parts of continental Europe these lampenflora biofilms can be a particular problem, for they can cause damage to prehistoric cave art.[20]

Although sometimes seen as unwelcome by the managers of tourist caves – there shouldn't be photosynthetic organisms growing deep in a cave where it should be dark, especially if they are damaging prehistoric art – looked at another way they are a fascinating illustration of microbial dispersal. Clearly photosynthetic organisms live outside where there is sunlight, and are managing to reach small islands of artificial light, deep in the cave, by some mix of air currents, flowing water or dispersal by humans or other animals such as bats. On a small scale in a show cave like Ingleborough, therefore, we appear to be seeing 'everything is everywhere but the environment selects' portrayed in green on the cave walls. Spores or other cells of tiny photosynthetic organisms are presumably everywhere within the cave, but only selected by the environment where there is artificial light and water allowing them to grow, reproduce and colonise that bit of rock surface. This goes not only for the cyanobacteria and diatoms, but for the mosses too – their spores are often of a similar size to the testate amoebae illustrated in multiple images throughout this book, and so are likely to be dispersed in a similar way to these microbes. Although show caves demonstrate this most dramatically, on a smaller scale you can see similar things in underpasses with artificial lights, and in the tunnels, dungeons and cellars of historic buildings such as castles (Figure 59). All that is needed is artificial lighting, and enough moisture to allow algae and mosses to grow in conditions that would otherwise be too dark for photosynthesis.

Water running into caves provides one obvious means for microorganisms to get into the cave system from outside. However, airborne dispersal seems very likely too. As described in Chapter 3, gravity is a very weak force for organisms the size of microbes, so they can 'float' in the air, only sinking downwards very slowly. Much of my own involvement with the 'everything is everywhere' debate, over the past few decades, has been with trying to understand the scale of long-distance microbial dispersal, usually using testate amoebae. This work has taken several approaches – such as the statistical analysis of large data sets (mainly comparing the species found in the Arctic and Antarctic), and mathematical models of the way airborne microbial-sized particles can be spread around the globe.

It follows from the 'everything is everywhere' idea that if you compare two similar habitats that are a long distance apart you should find many of the same species – as the similar environment will have selected them from the pool of globally dispersed microbes. With this idea in mind, several decades ago I compared the testate amoebae species found in the Antarctic and Arctic,[21] later with colleagues revisiting this analysis to include Tibet – a high, cold landscape

FIG 59. Lamp flora in a tunnel at Ashby-de-la-Zouch Castle, Leicestershire. As well as microbial biofilms a range of mosses and some small lichens grow around the lamps on the damp tunnel walls.

sometime referred to as 'the third pole'.[22] Many morphospecies of testates – that is, species characterised by what they look like using microscopy – can be found in both the Arctic and the Antarctic (that's rather like finding penguins or polar bears living in both polar areas). Many are also found in the cold conditions of Tibet. It turns out that the smaller the testate species, the more likely it is to be found at both poles, and in Tibet as well. This effect seems to start below a size of around 150 μm, but is particularly apparent for the really small species, below 20 μm.

These results are compatible with the 'everything is everywhere' idea, at least for things no larger than the smaller testate amoebae. However, there is a potential complication. The smaller the amoeba the less detail you can see under a microscope, so there is the possibility that so-called cryptic species are confusing things. That is, species that look the same under the microscope but in reality are different – though often it is only molecular analysis of their genetic material that can show the difference (Box 3). Such molecular analyses are however still relatively rare for polar testates. There is some relevant evidence for diatoms, where molecular evidence shows that some apparently widely distributed 'species' found in Antarctica are genetically quite distinct from individuals sampled from other parts of the world.[23] It's currently unclear if this is the case for most groups (for example, as I just pointed out, there is as yet relatively little molecular work on Antarctic testates).

In ecology, a complex question is not often going to be definitively answered by a single approach. More likely, several different approaches to the same problem will help converge on an answer. Another way of thinking about the

Box 3. Cryptic species

In the microworld the term *cryptic species* is used to refer to the situation where two or more distinct genetic lineages are effectively impossible to tell apart based on their morphology when viewed with a microscope. The term *pseudocryptic* is sometimes used for species that appear identical using a conventional light microscope, but can be differentiated using electron microscopy, or sometimes with light microscopy by using more elaborate staining techniques. However, such methods are impractical, or simply not available to many ecologists studying microbes collected from the field. With the rise of molecular methods over the last few decades, which can look directly at the microbes' genetics, there has been a large increase in the number of cryptic species identified.[24] The testate amoebae have produced several good examples of this problem, for example *Hyalosphenia papilio* and *Euglypha rotunda* – both of which are discussed in this chapter.

Cryptic species are not restricted to microorganisms, but are a particularly common problem in the microworld. A more familiar British example from the macroworld are the bats that used to be called the Common Pipistrelle *Pipistrellus pipistrellus*. In the 1990s people started to suspect that two almost identical-looking species might be hiding under this name, as different individuals used different echolocation signals. Further studies, including molecular analysis, confirmed that there are indeed two separate species, and the Soprano Pipistrelle *P. pygmaeus* was added to the list.[25] As a bat detector can tell the two apart based on their echolocation signals, using the terminology introduced at the start of this box they would best be described as pseudocryptic species.

'everything is everywhere' question is to use large computer-based models of the global atmosphere to investigate the distance airborne particles (i.e. microbes) may be able to disperse. For example, can they realistically blow between continents? That's one of the things I was doing in Switzerland in 2009 when I took the day off to botanise on Rochers-de-Naye. During my sabbatical visit to Lausanne I had access to much more powerful computers than I did back in Liverpool, and the right collaborators to make use of them.[26] What we did was to modify a large computer model that was designed to look at the dispersal of airborne pollution around the world to work for virtual microbes of a similar size to testate amoebae. We found that in our computer model, virtual microbes with a diameter of 20 μm or less could readily be blown between continents (e.g. across the Atlantic) but this became less likely as we made our virtual microbes larger. Above 60 μm, such long-distance dispersal did not occur – at least on the time scale of a single simulated year.

Our modelling study was unusual in trying to simulate distance travelled. Other similar studies modelling global atmospheric microbes had focused on simulating microbial concentration in the air using smaller microbes. Most had looked at bacteria (using cell diameters of around 1 μm), while some had been interested in fungal spores (using 5 μm diameters – although some fungal spores can be much larger than this). Unsurprisingly, these studies suggested that the highest densities of bacteria and fungal spores occur over the continents, but also suggested the possibility of global airborne dispersal, assuming the microbes can survive high in the atmosphere. This is an issue still not fully understood, and many will probably not survive due to high levels of ultraviolet light and a range of other stressors. However, such are the numbers in the air that even if only a very small percentage survives, that still gives very large numbers of live microbes arriving at distant locations.[27]

The results I have been describing are based on computer modelling studies; actually measuring long-distance dispersal of microbes in the air is obviously technically challenging. When we published our modelling study in 2012 we could only find around half a dozen studies convincingly showing movements of 2,000 km or more. However, technical improvements, especially in molecular methods, are starting to make such studies slightly easier. Recently, Rodó and colleagues found a wide range of bacteria and fungi in air collected from 3,000 m over Japan – including many species that can potentially cause illness in humans.[28] So far, such studies have tended to concentrate on bacteria and fungal spores, rather than the protists that are the main subject of this book. This is typical of much of microbiology, where the research funding tends to be concentrated in areas of medical, rather than environmental, interest.

Airborne dispersal is likely not the only way microbes disperse long distances. Think of the lamp flora in show caves like Ingleborough. As well as airborne dispersal, I suggested that human visitors, and other animals such as bats, might be bringing these photosynthetic microorganisms deep into the cave. One of the clear results from our modelling study was that, because of the global patterns of air circulation, it was very difficult for airborne microbes to cross the equator. However, large numbers of birds migrate between hemispheres annually, and in the modern world many humans do so too. Therefore it's likely that animals, as well as air currents, may be important in the long-distance dispersal of microbes. The potential role of birds is discussed in more detail later in this chapter in the context of algae moving between lakes, and I will return to the potential role of human dispersal in Chapter 9.

The approaches to addressing the 'everything is everywhere' question described in this section have been rather technical, relying on the statistical analysis of large data sets, computer models of microbes in the global atmosphere, or molecular data on microbes at high altitude collected from flights by research planes. But there is another approach that is much less technical, relying mainly on simple microscopy – a style of research used by nineteenth-century naturalists but still able to address cutting-edge questions today. A classic example of this approach is illustrated by a distinctive testate amoeba first described by Joseph Leidy in the nineteenth century.

THE LOST AMOEBA OF JOSEPH LEIDY

Searching for amoebae in his beloved Cedar Swamps in New Jersey, USA, in the 1870s, Joseph Leidy found a striking-looking testate amoeba, with strange projections which made it look a little like the test had arms. There were a few other nineteenth-century records of this species, along with two from the first half of the twentieth century – one published in 1952 from a site near Leidy's original location, and one in 1945 from southern Quebec in Canada. After that, this distinctive, unmistakable testate vanished from the scientific literature until 2011. I was one of the authors of the paper that announced the rediscovery of *Nebela ansata*,[29] recently moved to a new genus *Cornutheca*.[30] I will relate the background to this work in some detail – both because this species is one of the best examples we have of 'everything NOT being everywhere' and because it helps illustrate why many scientific research papers end up with a long list of authors. This latter point is something that seems rather odd about the scientific literature to people trained in many areas of the humanities, where academic articles still often have a single author.

In 2008 I bought a copy of Leidy's 1879 monograph, which includes a striking illustration of *Cornutheca ansata* (Figure 60). Intrigued, I searched the literature and realised that there were very few records of this species, with the most recent one I could find being 1952. This caused me to become curious about this testate, so I emailed Bob Booth at Lehigh University in Pennsylvania, as he was the nearest testate amoeba specialist to Leidy's original sites. He thought that the original sites had probably been lost to subsequent development, but later that year Bob and his collaborators collected moss samples from the New Jersey Pine Barrens as part of a wider study, and he searched through them looking for *C. ansata*. As he wrote in an email to me on 23 November 2008, 'I was beginning to think that we were not likely to find it, but on Friday I found a couple of individuals in one sample!' Shortly after that I received an email from Edward Mitchell in Switzerland to say that he and Thierry Heger, along with Canadian colleagues, had found *C. ansata* in Nova Scotia. I suggested Bob

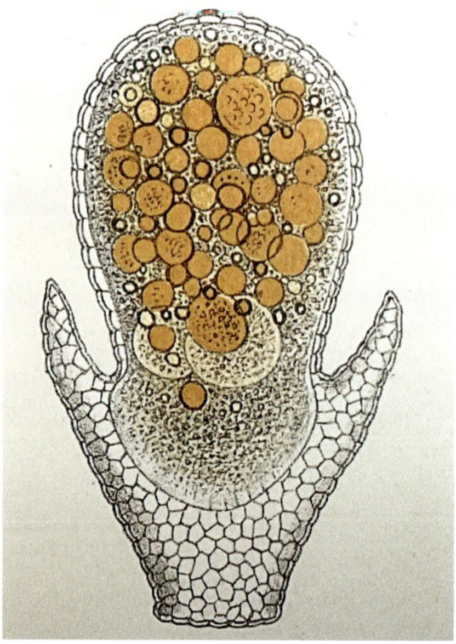

FIG 60. *Cornutheca ansata* (formerly *Nebela ansata*) as illustrated by Leidy (1879). He collected this specimen from bog-moss *Sphagnum* sp. in New Jersey, USA, in 1877. He described it as a 'well fed individual', noting that all the food particles concealed the nucleus from view. See also Figure 55.

and Edward should talk to each other, and they decided that rather than writing two separate research papers on the rediscovery of the species it was better to combine forces and collaborate on a single more substantial report. *C. ansata* was somewhat more common in the Nova Scotia samples than it was in Bob's from New Jersey, and it proved possible to extract DNA from the more plentiful Canadian samples for molecular analysis.

While the lab work was being done, serendipity continued to play a role. Yuri Mazei, from Penza State University in Russia, had been visiting the Natural History Museum in London and took the chance to visit Liverpool as a tourist early in 2010. I was at Liverpool John Moores University at the time, and he arranged to call in to talk testates. He had been at the Natural History Museum to look at the Penard collection. This comprises almost 1,000 microscope slides of amoebae sold to the museum at the start of the twentieth century by Eugène Penard.[31] Yuri had been looking at slides of testates in the genus *Difflugia*, but with a bit of time to spare he had browsed through the rest of the collection. When I mentioned the ongoing work on *C. ansata* he said he had seen an example in the Penard collection. This turned out to be a previously unpublished record from near Lakehurst in New Jersey, which had been collected by George Herbert Wailes and sent to Penard in Geneva. Chance had brought together records of *C. ansata*, both old and new, allowing this species to become a key case study in the debates over whether 'everything is everywhere'. In addition, we brought in Ralf Meisterfeld from Germany, one of the most knowledgeable testate taxonomists, who over a long career had assembled an extensive library of obscure and hard-to-find research papers on testate amoebae – ideal for tracking down the earlier records of *C. ansata*.

In our paper on this work we wrote that 'These data provide an unusually convincing case of a free-living microorganism with a very limited distribution'.[32] Why 'unusually convincing'? First, it's a morphologically very distinctive species, so even a non-specialist seeing it under the microscope is likely to recognise it as something interesting. It's found in a small number of peatlands down the east coast of North America, but not more widely across America, and not in Europe. This is important, as these are the best-studied regions of the world for testate amoebae. An absence from, for example, Africa or Australia would be harder to interpret, as these are regions that have been much less well studied for testates – so raising the question: is it truly absent or is it just that it hasn't been found? However, in North America and Europe something this distinctive should have been found if it was widely distributed. So this looks convincingly like a case where 'everything is not everywhere'.

But what about the idea that 'the environment selects'? Is it possible that we only find this testate at these few sites because it has rather specialised ecological requirements that are not found on other bogs across North America? To try and rule this out we carried out a statistical analysis of testate amoeba communities from 1,012 samples from similar habitats across North America and compared these with our two sites with *C. ansata*.[33] This suggested that, except for the presence of *C. ansata*, there was nothing special or unusual about these two sites. So it seems this is indeed a conclusive case of everything not being everywhere, where we can also rule out the absences being because the species has not been selected by the environment.

Because *C. ansata* seems to be restricted to a relatively small number of peatlands along the east coast of North America it's also an example of a microorganism that humans could plausibly make extinct, if we drained or otherwise modified these peatlands. The possibility of human-caused extinctions of free-living microbes is a topic I will return to in Chapter 9. One potential explanation for *C. ansata* being restricted to these bogs is that it's thought that the east coast was a location where peat bogs survived the last glacial maximum, whereas most North American peat bogs have formed since the ice sheets retreated. *C. ansata* is large for a microbe – usually well over 200 μm in length – and so, as described earlier in this chapter, unlikely to easily spread on the wind. Currently, our best guess is that it survived ice-age conditions at, or very near, its current sites and has been unable to spread much since.

I have described *Cornutheca ansata* in some detail, despite it not being a British species, because it is the best example I know of where a free-living protist appears to have a highly restricted distribution, with good evidence that this is not due to it having very specialist habitat requirements. However, there is a recently discovered example of something that seems similar from Britain and Ireland, which also has a Joseph Leidy connection because it involves a particularly distinctive form of his favourite testate species, *Hyalosphenia papilio*. As described in the previous chapter, recent molecular work has demonstrated that there are a number of distinct molecular types hidden within this 'species' – although they tend to look identical when viewed with either light or electron microscopes. However, Richard Payne noticed some very oddly shaped *H. papilio* from Cors Fochno, a peat bog on the Welsh coast (Figure 61). Richard died before he had the chance to complete a study of these, but what his colleagues found is fascinating. These testates looked so different from the standard *H. papilio* that they were described as a new subspecies called *H. papilio paynei* – in tribute to Richard. This is interesting, because Britain and Ireland (a few individuals of this distinctive subspecies have also been found at Annaghbeg peatland in the middle of Ireland) are relatively well studied for testates in bogs, but before Richard no one had noticed this distinctive form. There was enough material from Cors Fochno for molecular analysis, which confirmed that this new form differed from previously studied genetic lineages of *H. papilio*. It seemed to be particularly associated with the very wettest parts of the bog, and intriguingly the 'normal'-shaped *H. papilio* were unusually scarce at Cors Fochno. Unlike the case of *C. ansata*, the (currently) one Welsh and one Irish site for *H. papilio paynei* could not have been there in the last glaciation; the most plausible guess is that this distinctive subspecies has recently evolved somewhere on the Celtic fringe, since the ice retreated.

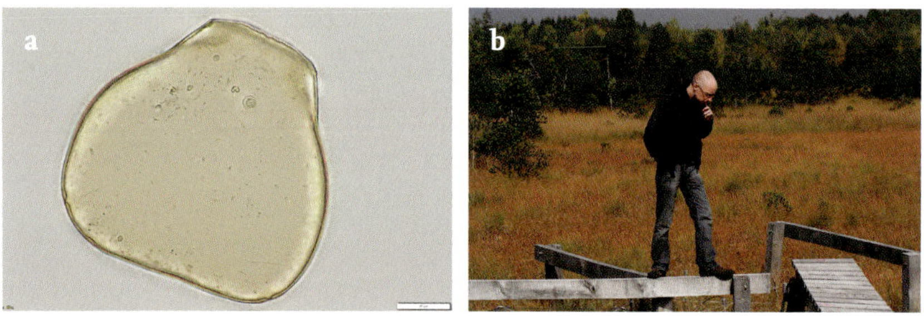

FIG 61. (a) The distinctive *Hyalosphenia papilio paynei*. Compare with the normal morphotypes shown in Figure 52. (b) The subspecies is named after the late Richard Payne, who first found this unusual testate at Cors Fochno (UK) – often referred to as Borth Bog in the older literature. Richard died in an avalanche in the Indian Himalaya before this testate was formally described by Duckert *et al.* (2023). The photograph shows him looking thoughtfully at the climate-change experiments on the peat bog at Forbonnet in the French Jura, in 2009. These experiments are described in Chapter 9. (a, Clément Duckert and Edward Mitchell)

Cornutheca ansata and *Hyalosphenia papilio paynei* are examples of what have come to be called 'flagship species' (or, in the latter case, subspecies). In the way microbial ecologists use the term, these are taxa that appear to provide clear examples of everything *not* being everywhere. However, it is another testate species that is the classic textbook example. This is *Apodera vas*, a species that is found widely in the southern hemisphere, but is missing from suitable habitat in Europe and North America (this species is discussed in some detail in the context of climate change in Chapter 9). A lesser-known example is *Bullinularia foissneri*, a large testate amoebae species known only from Mount Buffalo National Park in Victoria, Australia. In the original description of testates from this national park it was only identified to genus level, but later Ralf Meisterfeld described it as a new species following extensive study using scanning electron microscopy.[34] It is presumably more widely distributed than just this one national park, but there are very few testate amoebae studies from continental Australia. There is an obvious question: is it restricted to Australia – as so many larger organisms are – or has it been overlooked in samples from elsewhere? Ralf was at the time a leading testate taxonomist; perhaps many ecologists with an interest in this group would just assume they were looking at one of the more common species such as *B. indica* (especially if only using light microscopy), and so missed it.

Flagships have been identified not only among testate amoebae. Wilhelm Foissner – for whom *B. foissneri* was named to mark his sixtieth birthday – identified a number of flagship ciliate species. However, several of these have subsequently been found on other continents.[35] This illustrates a problem with the flagship species approach to the 'everything is everywhere' problem. A flagship which appears to show that everything is not everywhere may, a few years later, be found to be more widely dispersed. This however appears exceptionally unlikely in the case of *C. ansata*, as it is so clearly distinctive and missing from some of the best-studied peatlands which have been shown to contain the correct habitat for it. So there are some flagships which show that everything is not everywhere, but it's hard to know if these are just rare exceptions, or if many protists have limited ranges. And when we come to consider the distribution of protozoa and algae limited to freshwater habitats we encounter a particular problem, as to disperse between sites they have to cross often extensive areas of unsuitable habitat.

OF DUCKS AND ALGAE

Consider a lake such as Windermere (Figure 62). How could a microbe – such as a diatom or an amoeba – living in such a site move to a new lake? One obvious way

FIG 62. Windermere in the English Lake District. How could a microbe living in a lake such as this move to a new lake? To get to one of the other lakes in this region it would have to somehow cross the local hills. The easiest way to do this is to fly – but how can a microbe such as a diatom do this?

is through airborne dispersal, but it's not straightforward for a microbe to escape from the water into the air. One possibility is that they may use the wind. At wind speeds of around 20 km per hour wave tops start to break, with bubbles of white water being released into the air – taking microscopic plankton with them. Most of these just fall back into the lake, but a few will be wafted higher into the air, potentially allowing aquatic microbes to be blown long distances.[36] However, another possibility has already been briefly mentioned in this chapter, namely birds. A lake such as Windermere has plenty of ducks and other waterbirds: could they play a significant role in the dispersal of planktonic microbes?

The choice of Windermere as my example lake was not accidental. Some of the first good studies of the potential of waterbirds to disperse algae were carried out here in the 1970s by Kathleen Atkinson (Figure 63). Atkinson worked at the FBA at Windermere from 1961 to 1995, and on retirement she was awarded an MBE in the 1996 New Year's Honours. She trained as a botanist, and much of her work was on freshwater algae, but she was also a keen birdwatcher, and this led to a series of experiments, mainly during the early 1970s, on ducks and algal dispersal – although some of her colleagues at the time apparently thought this rather frivolous research.[37] Prior to her experiments, Vernon Proctor had published a short but influential paper in *Science* on viable algae he had found in the gut contents of migratory birds.[38] Atkinson cited Proctor's work in the first of her research papers on the topic. Like Proctor, she looked for algae in the guts of dead birds, which she described in a 1972 paper,[39] but most of her work was more experimental.

Both Proctor and Atkinson found what looked like viable algal cells in the guts of waterbirds – but that is some way short of showing that the cells can pass all the way through the bird and out in its faeces in a viable state. Dead waterbirds for dissection are, however, relatively easy to come by – either shot by hunters (as in Proctor's study), or in the case of Atkinson's work accidently drowned in gill nets used for fish sampling – so such birds make a useful starting point. However, Atkinson took things further in several experiments where she fed water rich in algae to captive waterbirds and then looked for viable algae in the faeces. The birds used in these experiments came from the Wildfowl & Wetlands Trust – with early work being carried out at Slimbridge, and later larger-scale experiments with birds housed at the FBA in Windermere. She found that her results were variable, with algae surviving gut passage better in Mallards *Anas platyrhynchos* than in the two species of diving duck she also studied – the Tufted Duck *Aythya fuligula* and the Australian White-eye *Aythya australis*. The diatom *Asterionella formosa* seemed particularly successful at still being viable after having passed all the way through a Mallard gut. She also found diatoms in the

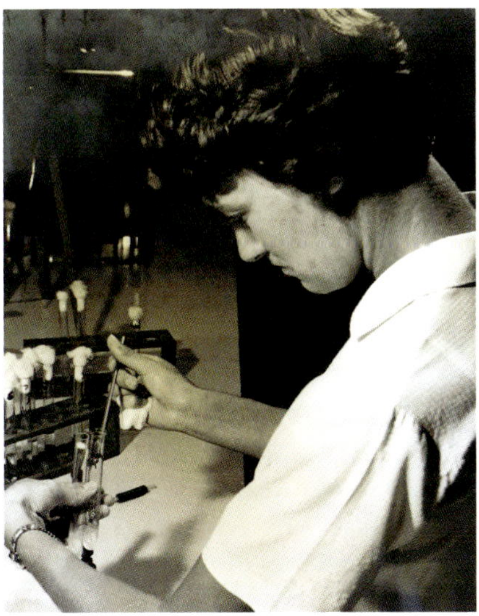

FIG 63. Kathleen Atkinson in the lab (most likely working on algal cultures). The picture is from a family scrapbook, photographer unknown. Atkinson published four papers on birds and algae between 1970 and 1980 – two in an ornithological journal (*Wildfowl*) and two in an algae journal (*British Phycological Journal*). Her final 1980 paper was the longest, and cites all her previous studies. She also has a link to the history of the New Naturalist series, supplying data on the birds of Windermere for Pearsall and Pennington's volume on the Lake District (1973). Some of her other work is briefly described in a note to Chapter 9. (Photo courtesy of Hilary Atkinson)

genus *Melosira* passing alive through ducks. She pointed out that although her studies had mixed success at showing algae surviving passing through birds' guts, because of the large numbers of such birds, many of which fly long distances, this could be a very important dispersal mechanism (Figure 64).

In 2016, unaware of Atkinson's early work, I also used Windermere as one of two sites for some pilot studies of waterbirds' abilities to disperse live algae. Rather than taking Atkinson's experimental approach with captive birds, with colleagues I collected Mute Swan *Cygnus olor* faeces from Windermere and Mallard faeces from Rostherne Mere in Cheshire. The Mallard samples contained viable-looking algae – especially the diatom *Pinnularia* sp., although while still green and obviously alive it didn't look particularly healthy. With my relatively simple attempts at culturing these I was unable to grow them to prove they were still viable. From the swan faeces I was able to culture cyanobacteria (Figure 65), but with my rather basic lab facilities I was unable to be sure that there had not been contamination from potentially airborne cyanobacteria. Because of these problems I never attempted to formally

FIG 64. Pink-footed Geese *Anser brachyrhynchus* at the Wildfowl & Wetlands Trust Martin Mere reserve in Lancashire. The large numbers of waterbirds such as these mean that even if the dispersal of viable algae is relatively unlikely, it will still happen on a regular basis because of the number of opportunities – think of each bird as a lottery ticket bought by an algal dispersal syndicate: the more tickets they buy the more likely they are to win (that is, colonise a new site).

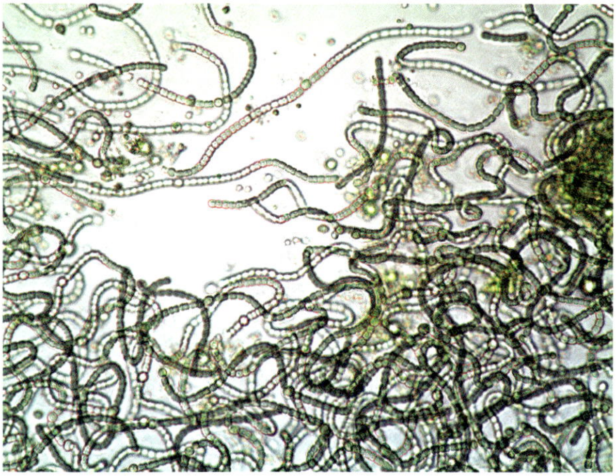

FIG 65. A tangle of cyanobacteria cultured from Mute Swan *Cygnus olor* faeces collected from Windermere. These are *Anabaena sensu lato* – this genus has recently been split into several different genera, based on molecular data (Kaštovský, 2024).

publish this work. Some of these are issues that Kathleen Atkinson must have faced too, although she was much more expert at culturing algal cells than I am.

Atkinson's experiments were pioneering, but few people followed up on her work; indeed, a large recent review of the role of waterbirds in the dispersal of plants and other organisms identified this as a still understudied area, except for studies on disease-causing microbes, which have had rather more attention.[40] Things may be starting to change, and one very recent study looked at ciliate protozoans from waterbird faeces collected in the Doñana National Park in southern Spain. They successfully cultured 39 morphospecies of ciliates from 20 faeces which had been collected from the field and then incubated in the laboratory – taking precautions to exclude the possibility that their cultures were being contaminated by airborne microbes.[41]

The other way waterbirds can disperse algae is externally on their feathers, but this is even less well studied than internal transport. One recent experiment that aimed to address this shortfall looked at the survival of diatoms on feathers from Mallards. Feathers were dipped in water containing the diatom *Nitzschia pusilla* and then subjected to a range of different relative humilities for periods of 10 minutes to three hours – before testing to see if viable diatoms survived on the feathers. The higher the humidity used, the better the diatoms survived.[42] A recent Chinese study has also shown cyanobacteria surviving for 90 hours or so on duck feathers, potentially allowing long-distance dispersal of the algae by waterbirds.[43] Although our current knowledge of the role of birds in moving algae between waterbodies is quite limited, it looks as if they may be potentially important. There is still much we don't know about the dispersal of free-living microbes.

MAKING SENSE OF MICROBIAL DIVERSITY

Over a century on, how should we view Penard's statement that he found 92 per cent of the then known global list of testate amoebae in his Lac Léman study? It's clear that the small size, and often colossal abundance, of many microbes makes long-distance dispersal more likely than it is for most larger organisms. But is everything really everywhere? Examples such as *Cornutheca ansata* show that it's not literally true, but could it still be correct in the majority of cases? One complicating factor is that now we can look at a microbe's genetic makeup, rather than just what it looks like under a microscope, and we are in some cases finding cryptic species. Examples of testate amoebae discussed above include *Euglypha rotunda* (Figure 57) and *Hyalosphenia papilio*. But this is not a problem restricted to testates. For example, Jan Kaštovský has recently estimated that of the 466 new species of cyanobacteria described since the start of the twenty-first century 296 were cryptic – that is, they can only be identified by DNA sequencing and are indistinguishable by microscopy.[44]

One possibility is that we have not been comparing like with like when we contrast the distribution of species of larger organisms with those of microbes, especially microbial morphospecies. In some of the cases of cryptic species, what we have been calling a microbial species may be more equivalent to a genus, or even family, in more familiar larger organisms. This would make the much larger ranges of many microbial species less surprising, and more in line with what we are used to seeing in larger organisms.[45] Consider a botanical example. The Yellow Horned-poppy *Glaucium flavum* (Figure 66) is a specialist of coastal shingle and sand found from Britain to the Mediterranean. This Mediterranean distribution means that while mainly a European species it is technically native to three continents – just getting into Africa and western Asia. However, the genus it belongs to is found much more widely across Asia, and its family is found on

OPPOSITE: **FIG 66.** Yellow Horned-poppy *Glaucium flavum*, a species common on coastal shingle, but one that can also sometimes be found on other disturbed sites. (a) On the Norfolk coast near Cley, and (b) growing at the back of a beach at Palaiókhora, Crete. The species' Mediterranean distribution means that it just gets into Africa and Asia as a native species (it is naturalised more widely, including in North America). However, other species in this genus are found more extensively across southwest and central Asia. The poppy family (Papaveraceae) is distributed more widely again, across Eurasia and North America, with a more limited distribution in South America and Africa (Mabberley, 2017; Christenhusz *et al.*, 2018). Possibly many microbial 'species' are better thought of as widely dispersed genera which, over time, will have evolved into multiple species across their global range.

all continents except Australasia and Antarctica. If we could only recognise this plant at genus or family level we would assume that it was a 'species' with a much larger native distribution than is actually the case. Increasing molecular data should help here, so that cryptic species become less cryptic and more visible. However, it's still not obvious exactly how you define a species of microbe based on molecular data. At the very least such data do demonstrate a much greater variation in some apparently widely distributed morphospecies. Indeed, even before molecular (that is, gene-based) data became available, earlier biochemical methods were suggesting to some scientists that we might have overestimated the extent to which everything was everywhere.[46]

While some very good protist researchers still argue that 'everything is everywhere but the environment selects' is a very good approximation of reality,[47] the current view among most people working in the area is rather more nuanced. Some species of microbe are indeed very widely distributed, but others have much more limited distributions (*C. ansata* is one of the more extreme, and convincing, examples of this). Recently, Eveline Pinseel and colleagues have suggested that this less polarised view allows us now to move on to new questions – such as trying to understand why *some* protists appear to have near-global distributions, while *other*s are more limited.[48] Can we start to understand what it is about the natural histories of these different microbes that lead to these different biogeographical patterns. For example, when is wind, or bird, dispersal important, and for which types of protist? This is more complex than a simple story of everything being everywhere, with the environment then selecting, but it makes sense in the context of my own work in this area. Since I first started publishing research on large-scale protist biogeography in the 1980s, my work has been cited both in support for everything being everywhere, and as showing that this is clearly not the case – as different studies have produced results that were sometimes more compatible with one extreme than the other. This is what I would expect in a complex world where both extremes turn out to be oversimplifications. However, these past simplifications were influential in highlighting the importance of understanding microbial dispersal, and in pushing people to further studies to try and clarify the actual situation.

CHAPTER 6

Sex and Death

Late spring, and the dead wood of the veteran oaks stands out under a clear blue sky, contrasting with the deepening green of the new leaves. I haven't brought a tripod with me, and a large fallen tree trunk serves to steady my camera as I photograph the oaks among the emerging bright green of early-season bracken fronds (Figure 67). Although I am only 10 minutes or so from the main car park at the National Trust's Calke Abbey, as I am away from the paths there are few people around. Most of the visitors are concentrated around the buildings, now seen by many as synonymous with the activities of the Trust. Following the First World War the large country houses of Britain became increasingly unaffordable as private homes, and during the 1930s the National Trust started to try and salvage some of them through its Country House Scheme.[1] In so doing it ended up acquiring many fine, old, parkland trees such as the ones in front of me – likely more to the taste of many of the Trust's founders, given their nature-conservation interests, than the large historic houses that came with the parks. Calke itself wasn't acquired by the Trust until the 1980s, and the buildings are now maintained as an example of a country house in mid-twentieth-century decline – I have been told that the gardeners have long discussions about exactly how many weeds to leave around the buildings to give the impression of an estate that can no longer afford enough garden staff, and is quite literally 'going to seed'.

These old oaks raise interesting questions. Some are clearly already dead, while others seem aged but alive with varying amounts of dead wood. The most long-lived British oaks can be something like 1,000 years old and can, like any organism, die of accident or disease. But do oaks senesce in the same way that naturalists, and other mammals, clearly do? At the other end of the scale

FIG 67. Dead and living Pedunculate Oaks *Quercus robur* in late spring with emerging Bracken *Pteridium aquilinum* fronds in a woodland clearing in the grounds of Calke Abbey, in southern Derbyshire. The parkland at Calke is 'especially well-endowed with mature trees' (Willmot & Moyes, 2015). The oldest British oaks can be 800–900 years old (Farjon, 2023); note the stag-headed tree on the right (with dead wood standing proud of the canopy), discussed in more detail in the main text.

of lifespans, many microbes often have very short lives. Can these organisms senesce? What would old age look like in a protozoan or diatom?

And what about sex? If the novels or period dramas set in country houses are to be believed, sex is intertwined with death in any good plot. Walking down the modest hill from the woodland clearing through more open parkland brings me to the big house. Here sex is extravagantly on show in the grassland viewed across the ha-ha, the ground bright yellow with buttercups all advertising to pollinating insects (Figure 68). However, the extent to which microbes engage in sexual reproduction is still uncertain, with modern molecular methods only now uncovering protist sex in some groups traditionally assumed to be asexual. After all, it was only around the end of the seventeenth century that botanists really started to understand that flowers were all about sex,[2] and buttercups are much easier to study than diatoms or testate amoebae.

FIG 68. A spring display of sexual advertising at Calke Abbey – viewed over the wall of the ha-ha with the main house in the background. The most prominent flowers are buttercups (mainly Bulbous Buttercup *Ranunculus bulbosus*) and Hawthorn *Crataegus monogyna*. Both buttercups and hawthorns are pollinated by a wide range of insects (Proctor *et al.*, 1996).

Some biologists now see important links between senescence and sex in an organism's life history– and if so, protists can potentially help us understand these two classic themes of life and literature. Back in the 1980s the ecologist and evolutionary biologist Graham Bell developed these ideas in a book called *Sex and Death in Protozoa* – this chapter is in part inspired by his approach.[3]

SENESCENCE: IN TREES, HUMANS AND PROTISTS

Although the title of Bell's book referred to 'sex and death' it was really senescence, rather than death, that he was interested in. Biologists tend to define senescence as a decline in an organism's performance with age – for a tree, that might be a decrease in its ability to produce seeds, or, for a protist, the rate at which it can divide to produce new cells. For Shakespeare in *King Lear* it was marked by mental decline, the ageing king fretting that 'I fear I am not in my perfect mind.

Methinks I should know you and know this man, yet I am doubtful'.[4] In humans you also see a decline in athletic performance – for example, the world record for the 100-metre sprint by a centenarian is no faster than the time most (non-athlete) 20-year-olds would take to run the same distance. We are so used to seeing older people decline and start to malfunction, that it just seems a natural part of life. However, if you think about it a bit more deeply there is an interesting biological question here – why can't the body fix the infirmities that accrue with age? After all, you start life as a single cell. If biology can create something as complex as a human from one cell, surely it should be able to keep an older individual healthy, fixing bits as they go wrong. Despite this feeling that it should be fixable, there is a large body of theory suggesting that there are good evolutionary reasons why this is not the case, and that senescence is something we should expect to see (Box 4). Some of these ideas don't apply to single-celled organisms, but the key ones do, and they imply that senescence is likely to occur even in microbes.

It's a common observation that senescence affects both humans and our pets; it has been much less obvious if similar things happen to plants, and senescence in microbes has been even more obscure. A key way that plants, such as the Calke Abbey oaks, differ from the animals we are most familiar with is that plants are usually modular organisms – adding new structures such as leaves, roots and branches in a way that would be very surprising to see in most animals. Imagine sprouting extra limbs, or heads, over the course of your life. Aljos Farjon describes this well for the case of old oak trees:

> *This compartmentalisation, where large areas of the tree are dead but completely new parts are alive, is characteristic of tree growth. Whereas an animal replaces dead cells, a tree makes new cells in another place.*[5]

This allows trees potentially to restructure themselves as they get older. Well-known examples of this are so-called stag-headed trees (Figure 67). In the past, trees with such projecting dead branches were considered to be dying, but they are now often viewed as trees that are rearranging their crown, perhaps in response to stress from a past event such as a drought. Can they perhaps, by processes such as this, avoid senescence – redesigning and reinvigorating themselves over the years? If so, while trees (and other plants) will eventually die of some accident or other, they don't die of old age in the way we are familiar with. As Oliver Rackham wrote, 'oaks are not immortal: they die at random from unknown causes.'[6]

Is this actually the case? The short answer, as is so often the case in this book, is that the jury is still out. There are good studies that fail to find evidence for

Box 4. Why senescence? An evolutionary view

The key insights underlying our current evolutionary understanding of senescence are mainly due to Peter Medawar in the late 1940s and early 1950s (although others, especially J. B. S. Haldane and R. A. Fisher, had developed part of the answer somewhat earlier). Medawar realised that the longer you live the more chance there is that you will suffer some fatal accident, and because of this few organisms manage to live to an old age in the wild. This means that natural selection has little effect on things that happen later in life, as most individuals have done all their reproduction by then – indeed, many of them are already dead. A mutation that reduces an organism's ability to function when young will obviously tend to be selected out of the population. But a similar mutation that only affects older individuals will be almost invisible to natural selection and so not removed.

There is an additional related reason why natural selection doesn't act to fix the problem of senescence. Imagine a mutation that increases an organism's ability to breed early in life, but at the cost of increased problems later. As most individuals with this mutation will likely have died for other reasons by the time these late-acting problems set in, natural selection will see the early-life advantages, not the late-life problems. So a gene that increases your late-life problems will be actively selected for because of its benefits to younger individuals. Medawar also realised that this was potentially an important part of the evolutionary argument, but didn't develop this aspect of the theory (his main research was in immunology). Others, notably George Williams and Bill (W. D.) Hamilton, developed these ideas over the decade following Medawar's publications. The details of this idea have been the subject of much recent debate by theoretically inclined biologists, but it is clearly applicable in at least some cases.

In addition to these ideas there are other theoretical approaches to senescence that only apply to organisms that have separate reproductive and body cells – which is obviously not the case for the single-celled protists that are the focus of this book.

Accessible introductions to the evolution of senescence include Jonathan Silvertown's *The Long and the Short of It* (2013), and the first chapter of *Big Questions in Ecology and Evolution* (Sherratt & Wilkinson, 2009). These also provide more detailed descriptions of, and references to, the classic work by Medawar, Williams, Hamilton and others. A recent article by de Vries *et al.* on 'Extrinsic mortality and senescence: a guide for the perplexed' (2023) introduces some of the more recent theoretical debates on when these ideas are applicable, or when they are not.

senescence in plants. Consider the Bristlecone Pines *Pinus longaeva* found in a few of the mountains of California and nearby states. Some of the Californian examples reach ages of around 5,000 years, but don't reach any great size. Indeed, the oldest known specimen was dated by ring counts, after it was cut down, to 4,900 years – but at the time it wasn't considered to be an especially old tree as it was not particularly large.[7] Despite their age, these pines apparently don't show any sign of age-related decreases in pollen viability or seed germination.[8] And it's not just in trees that scientists have struggled to find evidence of senescence. *Borderea pyrenaica* is a long-lived herb found in the Pyrenees, which can reach ages of 300 years or so.[9] Individual plants can be aged using scars on the tuber where previous years' stems were attached, and there was no evidence of declines in growth or reproductive ability with age.

However, evidence of senescence is increasingly being found in many plants: for example, a huge study by over 60 scientists looked for senescence in 597 species of tree from around the world.[10] They used tree-trunk girth as an indicator of age, and found evidence for senescence in 80 per cent of the species they looked at. Interestingly, many of the species where they couldn't find an effect of senescence were those in which large (and so old) individuals were rare in their data set, suggesting that the apparent lack of senescence might be due to limited data. They also point out other complications with senescence in trees. For example, coppicing often seems to at least partly reset the clock, increasing the longevity of many trees.

What about senescence in the microworld? *Hydra* is an interesting organism for trying to answer this question. It's a multicellular, borderline microscopic, animal that can be found on pond plants and similar habitats (Figure 69). They can be a couple of millimetres long, so of a size where their basic tentacled

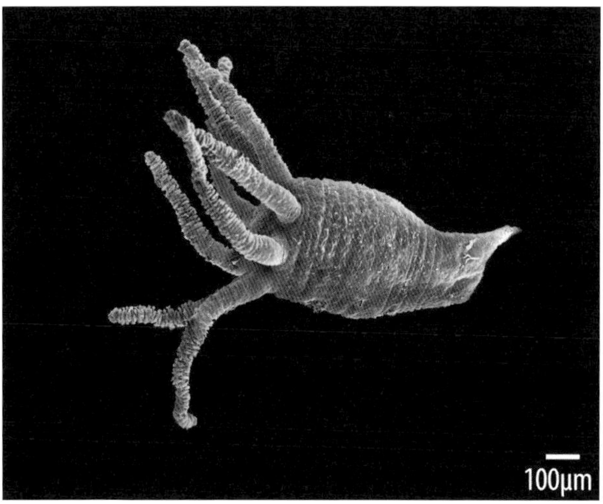

FIG 69. Scanning electron micrograph of Brown Hydra *Hydra oligactis* – a common pond species. As described in the text, there is little evidence for senescence in *Hydra* species (Dańko *et al.*, 2015). (Alex Hyde)

structure can be seen with a hand lens or low-powered microscope, and are in the same phylum as the more familiar sea anemones and jellyfish. They have relatively few cell types, and reproduce mainly by budding – with occasional sexual reproduction. Currently there is little evidence for senescence in *Hydra*, with neither an increase in the likelihood of death, nor a decrease in the rate of budding, in experiments that have run for four years or so. However, as with the more familiar trees discussed above, there is the possibility that senescence has been missed because not enough really old animals have been included in these studies. Perhaps four years is just not long enough to find convincing senescence in *Hydra*, but at the very least it seems these animals senesce very slowly.

What about senescence in the single-celled protists that are the main subject of this book? As discussed later in this chapter, these often reproduce mainly asexually, with sometimes occasional sexual reproduction, as in the case of *Hydra*. A key reason to wonder if senescence applies to protists is that asexual reproduction is common, and involves one cell dividing into two new individuals – albeit clones, genetically identical individuals (Figure 70). Could this lead to potentially immortal lineages of such clones? It seems logical that these two divided cells must start life identical to each other. However, this may not be the case, and even in dividing bacteria there is evidence that one cell may inherit older cell contents than another. If so, two cells, the product of a single cell division, may effectively start life with different ages.[11]

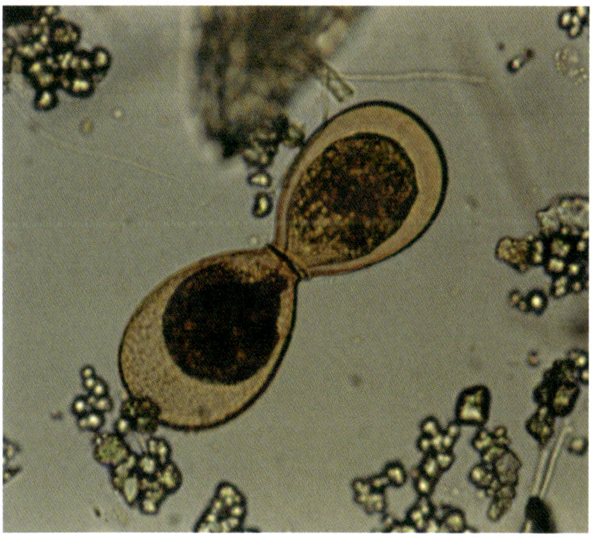

FIG 70. Asexual division in the testate amoeba *Nebela tincta*. One cell stays in the original shell while the other moves into a newly constructed shell. Here is one way in which the two new cells may not have the same 'age' – one has a brand-new shell while the other has inherited the old shell, with the potential for damage from past wear and tear. (Edward Mitchell)

In his 1988 book Graham Bell collated a large number of experiments where protozoa had been kept in captivity for many years to try and establish if these longstanding clonal lineages showed senescence over time. Many of the experiments had been carried out in the early twentieth century, when scientists seem to have had more time and patience for such long-running studies. His general conclusion was that over time these clonal populations often showed a decline in the rate of reproduction (i.e. cell division), which he interpreted as senescence in the clone. You can plausibly think of a population of clonal single-celled organisms as genetically a single individual. This seems a bit counterintuitive for an animal like us, but in fact many plant species show a similar population structure – one classic textbook example is a Banyan *Ficus benghalensis* in India which is over 200 years old but comprises a small wood of interlinked clonal tree trunks covering over 1.5 hectares.[12] If protozoan clones do senesce, then this should lead to microbial clonal populations becoming extinct on a time scale of decades, as seen in the lab experiments. But this is clearly not the case, as protozoa have a very long geological history (see Chapter 2). Bell suggested that sexual reproduction provided a key part of the explanation.

SEX, DIATOMS AND MAKING CHAIRS

At its simplest, sexual reproduction in eukaryotes involves combining genetic material from the parents so that the offspring's genetics is a blend of that of their parents – rather than being identical to the single parent as is the case in asexual reproduction (give or take the odd mutation). There is an obvious problem with this: what stops the offspring having twice the amount of genetic material found in each parent, with additional increases for each succeeding generation? The answer is the process of meiosis, where the amount of genetic material is halved before two lots of genetically half-sized cells (gametes: sperm and egg in mammals such as humans) combine to produce an offspring with a full genetic complement. This means that one way of trying to find sexual reproduction in a protist is to look for evidence of meiosis. Advances in molecular methods have made this much easier in recent years as people can look for the genes associated with this process.

Diatoms are a nice example of one way in which sexual reproduction can be important for the long-term survival of a protist population. Much of diatom reproduction is asexual, and involves the separation of the valves that make up the silica case (frustule). This gives two valves of different sizes that act as

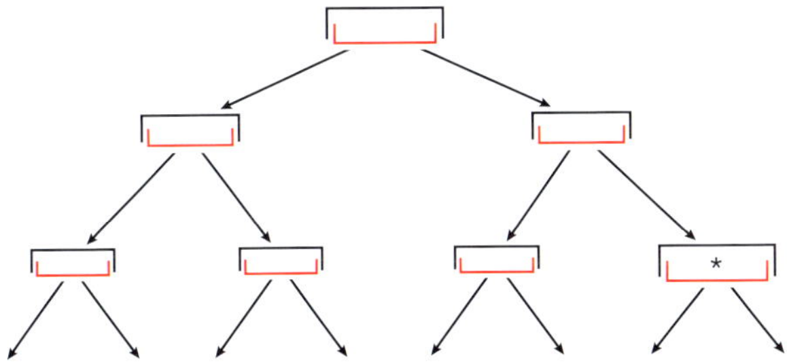

FIG 71. A diagrammatic representation of the progressive decrease in diatom size due to asexual reproduction. After cell division, each valve develops a new smaller base (shown in red), and this leads to a decrease in average size over time – in the bottom row of four diatoms only the one on the far right (marked *) is the same size as the original diatom. The extent to which this special case of size decrease in diatoms can be considered an example of senescence is controversial (Andrews, 2017). (Figure based on Bonner, 2013)

templates for new valves to form, with each old valve forming the larger valve of the new diatom. This means that at every asexual division one of the new diatoms will end up smaller than its parent. The result is that over time the population will be composed of smaller and smaller individuals (Figure 71). This potentially ever-shrinking diatom size, which may last for many months in culture, is corrected by occasional sexual reproduction. When this happens, two cells produce gametes which fuse to produce a new individual containing genes from both parents. Although the exact details differ between centric (radially symmetrical) and pennate (bilaterally symmetrical) diatoms, the result is an individual with a new frustule comprising two new valves, which is restored to the original size.[13]

There is clearly more to sexual reproduction than just preventing diatoms vanishing through size reduction. In his 1988 book Graham Bell suggested a furniture-making metaphor to help explain the importance of sex for protists and other organisms. Few biologists have taken notice of this metaphor; however, I agree with Matthias Galipaud and Hanna Kokko that this is a useful way of highlighting the problem and worth taking seriously.[14] Consider a chair maker who is in the business of producing lots of chairs but is not allowed to keep the original design blueprint – be that a drawing or a model chair – to refer to. The nearest they are allowed to a blueprint is to use the last chair they made as a model of how the next chair should look. This is analogous to asexual reproduction, and over time minor errors in chair construction will start to add up, making the chair less useful for sitting on. For example, in the diatom case the chair would become ever smaller over time – starting as a useful chair, but ending up looking more like doll's-house furniture as ever smaller frustules are used as models for making the next generation. The alternative approach is that the chair maker produces a large number of different chairs cobbled together out of assorted chair parts and these are then subjected to some test of functionality – for example being sat on by many people – with only the chairs that 'work' being put on sale. This is analogous to sexual reproduction, which mixes 'chair making' instructions from the parents, with each 'offspring' chair being different. In the first case, the functionality of each generation of chairs will necessarily decline over time, with the rate of decline depending on the accuracy and skills of the chair maker (an idea often referred to as 'Muller's ratchet' – after Hermann Muller, who did much to develop these ideas in the mid-twentieth century). In the second case, selection by chair users will remove defective chairs from the population, rather than them surviving to be models for next-generation chairs. Both approaches have advantages, and disadvantages – the second (sexual) approach is very wasteful in producing lots of defective furniture, but does provide a mechanism to prevent the long-term decline in chair function.

The avoidance of Muller's ratchet is not the only potential benefit from sexual reproduction. One way to think about the advantages from sex is to use the title of the 1966 spaghetti Western *The Good, the Bad and the Ugly* as an aide memoir for the different theoretical approaches.[15] 'The bad' is Muller's ratchet – the accumulation of damaging mutations over time in asexual reproduction. 'The good' is the idea that sex produces genetic diversity which may include new useful mutations, and potentially various beneficial mutations can be brought together in one offspring as they inherit genes from both parents. What about 'the ugly'? These are parasites, which in this context include entities such as viruses and bacteria, which can cause serious infections. The idea is that such parasites may be a particular problem for asexually produced clones. As these are genetically identical – give or take the odd mutation – it should be relatively easy for the parasites to evolve adaptations for successfully attacking these organisms. The genetic variation produced by sex leads to individuals with *rare* genetic makeups which should be harder for parasites to successfully target. These parasite-based arguments for the importance of sex are largely associated with work by W. D. Hamilton, starting in the 1970s – although, as with the ideas on the evolution of senescence, this was building on earlier speculations by J. B. S. Haldane.[16]

The reason why sexual reproduction attracted the interest of many of the most talented theoretical biologists of the twentieth century is that as well as advantages, there are some obvious disadvantages to sex too. One of the most famous of these disadvantages has become known as 'the two-fold cost of sex', an idea particularly associated with John Maynard Smith. The problem is that if you reproduce asexually then 100 per cent of your genes are found in your offspring, while in sexual reproduction it is only 50 per cent. This should favour asexual over sexual reproduction, as more of your genes get into the next generation – so benefits from sex need to be large enough to outweigh this disadvantage. In addition, imagine a reproducing organism, such as an algal cell. If it is in a position to reproduce then it must have what it takes to survive in that environment – asexual reproduction preserves this advantageous mix of genes while sexual reproduction risks producing something less well suited to the current environment. This may be an important reason why many protists seem to only indulge in occasional sex, with most of their reproduction being asexual, as in the diatom example. Asexual reproduction allows you to produce lots of diatoms with the currently successful mix of genes when the going is good, allowing rapid population expansion.

SEX IN THE MICROWORLD

My involvement with biological research started around the mid-1980s, and my impression is that the view of most biologists over the last 40 years or so has been that absence of sex in multicellular organisms is very unusual. Indeed, groups with a long evolutionary record where sex appears to be absent have usually been considered very odd and in need of explanation – sometimes called 'ancient asexual scandals'. Protists have been seen as more often asexual, and where sex is present it's been thought to often be only occasional, most individuals being the product of asexual reproduction – although there are exceptions: for example, sex has long been known in many diatoms, ciliates (Figure 72) and green algae.[17] However, things are starting to change, with evidence of at least occasional sex being found much more widely.

Amoebae make an interesting example (remember that not all amoebae are closely related to each other – it's a term that describes an organism's appearance and ecology, not its evolutionary history). Some amoebae are clearly sexual, namely the slime moulds described in more detail in the next chapter in the context of multicellularity. Even here it was not until the 1970s that sexuality was properly demonstrated in the cellular slime moulds. Sex has been known for longer in the plasmodial slime moulds, and here things are particularly complicated with up to 13 different mating types (the equivalent of different sexes). One thing the microworld shows is that the simple two-sex situation we are familiar with from larger organisms is not the only possibility. Ciliates also have multiple mating types (effectively many different sexes – Figure 72). However, in most amoebae evidence for sex has been much harder to find, and for the classic *Amoeba proteus* there is still no evidence for sexual reproduction.

FIG 72. Sex (conjugation) in the freshwater ciliate *Coleps viridis*. Sexual reproduction is widespread in ciliates, but there are some groups of ciliates where sexuality has still never been recorded (Esteban & Fenchel, 2020). In conjugation in ciliates, two cells of compatible mating types attach to each other, as shown here; this may last for many hours. They swap genetic material before separating (Margulis & Chapman, 2009). (Genoveva Esteban)

Modern molecular analysis has changed how we view sex in amoebae.[18] For example, there is now evidence that even in the groups of amoebae where we haven't (yet?) found evidence of sexual reproduction, they were sexual in the past. It looks as if asexual reproduction has evolved relatively recently in these groups. This is a familiar pattern from many plant groups, such as dandelions (Figure 73), where there is a mass of hard-to-identify asexual 'microspecies'. In an evolutionary context such groups are often relatively short-lived. They gain short-term advantages from avoiding the costs of sex, but long term their lack of sexuality makes them more prone to extinction, because they lack the advantages that come from some combination of the good, the bad and the ugly. Molecular evidence also suggests that the last eukaryote common ancestor (LECA – see Chapter 2) was sexual. If this is correct, then where sex is absent it has been lost at some point during that organism's evolution.[19]

There is one example of a group that is both asexual and has existed for at least tens of millions of years, that features very prominently in discussions of the evolution of sex – namely, the bdelloid rotifers. These are tiny animals which sit just on the edge of the microworld, being a millimetre or so long – so as with *Hydra* needing at least low-power magnification for you to start to see any detail. The bdelloid rotifers suggest some explanations for the long-term survival of asexual populations that are likely also relevant to the protists that form the main topic of this book. Because they are the classic example of an 'ancient asexual

FIG 73. Despite their flowers and copious seed production most dandelions *Taraxacum* spp. are asexual. That is why they are so difficult to identify to species level. Currently there are 248 microspecies (effectively asexual clones) known from Britain (Stace, 2019).

scandal' they have been very well studied, and no one has found any evidence for sexual reproduction. Many people have looked, as sex in these rotifers would be big news. The combination of the good, the bad and the ugly should make their survival over geological lengths of time unlikely, so what is going on?

Research in the first decade of this century suggested two potential explanations. First, molecular analysis has shown that these rotifers have experienced a significant amount of horizontal gene transfer.[20] That is, they can acquire genes from the environment, not just from a parent. This is unusual in animals, and more typical of prokaryotes such as bacteria. This potentially gives them a way of gaining genetic diversity without having to rely on sex. These rotifers can completely dry out and then become reanimated when water returns, a topic discussed in more detail below. It's plausible that it may be during such changes that much of their horizontal gene acquisition takes place. This ability to completely dry out is also fundamental to the other potential mechanism for sidestepping the need for sex – the 'ugly' problem of parasites. Small dried-out rotifers can easily be dispersed by wind, through blowing away as dust. Experiments in a wind tunnel, using bdelloid rotifers infected with fungal parasites, showed that new populations formed from wind-dispersed dried-out animals had often rid themselves of the parasites.[21] The drying killed the fungus. This gives a mechanism of escaping parasites which doesn't rely on sex producing genetically rare types. As many protists can also survive completely drying out and may often be wind-dispersed while dry, this may be an explanation that applies more widely than to rotifers. The ability to dry out raises all sorts of interesting questions, about the nature of death as well as sex.

DEATH AND ZOMBIES

Philosophers wondering about the nature of death are prone to inventing thought experiments along the lines of 'If I am decapitated, but a scientist rushes in announcing a new technique of sewing heads back on, as they now do for hands, am I brought back from the dead, or did I never die?'[22] However, we don't need such imagined scenarios to confront the difficulties of defining death; the microworld is full of real life-and-death examples. The dried-out rotifers in the experiments just described show no measurable sign of life (i.e. no measurable metabolism), and similarly sized tardigrades are even more famous for surviving hostile conditions in a dried-out state, but resuming living once they have been rewetted (Figure 74). As Anthony Flemming wrote in 2008, 'Tardigrades can be desiccated, irradiated, asphyxiated, poisoned, put in a vacuum, frozen to almost absolute zero, and yet survive.'[23] To survive the more extreme of these assaults they have to be in a dried-out state, apparently dead. The technical term for this state is anhydrobiosis; it's an ambiguous state sometimes described as 'latent life', but once rewetted and metabolising again the tardigrade seems very like the popular conception of a zombie – a reanimated corpse.

This hard-to-define boundary between life and death escapes from the microworld into visible natural history too; some mosses can come back from complete drying out, as can a very few seed plants, but it is the lichens that are spectacularly good at this. When this happens with a lichen, it's not just the fungus that is reanimated on adding water: so too are its photosynthetic symbionts (often green algae), along with potentially a range of other microbes that can form part of the lichen symbiosis. Most dramatically, some lichens have survived being exposed to the hostile conditions of space. In particular the elegant sunburst lichen *Rusavskia elegans* (formerly *Xanthoria elegans*), a globally widespread species found in Britain, survived for 18 months, in its dried-out form, exposed on the outside of the International Space Station. During this time it was dead – if your definition of life involves some sign of active metabolism. However, on return to Earth, and rewetting, the fungus came back to life in 84 per cent of cases, while its green algal symbiont reanimated in 71 per cent of cases.[24]

FIG 74. Tardigrades are famous for the ability, of at least some species, to survive almost any degree of hostile conditions in a dried-out form. They have even been reanimated after being exposed to the vacuum of space (Jönsson *et al.*, 2008). In the context of the ideas discussed in this chapter, it's of interest that tardigrades seem to be mainly sexual – although asexual and hermaphrodite tardigrades are also known (Flemming, 2008). This is a false-coloured SEM image of a tardigrade in the family Macrobiotidae found in damp moss growing on a house roof in Derbyshire. Damp/wet moss is a good habitat to search for tardigrades, which can be seen with relatively low-power microscopes, as the largest species are a little over 1 mm long (see Appendix). (Alex Hyde)

It is important that the lichen is fully dried up, vitrified in a glass-like state, as wet metabolically active lichens will not survive space travel.[25] The ability to survive dormant (or dead?) in space has made this lichen a favourite experimental organism of astrobiologists interested in the possibility of life on planets other than the Earth. It also allows lichens to colonise many habitats on Earth where they will regularly dry out (Figure 75). Similar behaviour is seen in a wide range of amoebae, which can sit out hostile conditions in an encysted state. Here the amoeba forms a spore-like structure which can sit there apparently doing nothing until better conditions return and it becomes reanimated. There is still much we don't know about encystment in amoebae. As Ascel Samba-Louaka phrased it in a recent short review article on the topic, there are 'so many blind spots to cover'.[26]

FIG 75. Lichens on a rocky shore on the Scottish island of Mull. The band of black rock is due to the lichen *Hydropunctaria maura*, often called the Black Tar Lichen, as it looks a bit like oil pollution on the rock (formerly *Verrucaria maura*; there may be several cryptic species hiding under this name). The yellow lichens are *Xanthoria parietina* (sometimes called 'Common Sunburst Lichen' or 'Common Wall Lichen'). The grey crust is mainly *Ochrolechia parella*, and the bushy grey lichen is sea ivory *Ramalina* sp. This habitat switches from very wet to totally dried out (on a sunny day), making the lichens' ability to survive drying out very important.

However impressive lichens, tardigrades and amoebae are, the real champions in sitting around inactive, before coming back to life, are found among the bacteria. Although this book focuses particularly on the protists, the really long-'lived' bacterial spores are worth a brief mention. Bacteria prompt us to ask the question, just how long can a microbe survive in an undead condition in rocks? The first person to have a go at answering this was Charles B. Lipman in the 1920s.[27] He looked for, and extracted, bacteria from rocks many hundreds of millions of years old and successfully cultured some of these in the lab. This is both a really interesting question to ask, and technically a very difficult thing to do.

One reason Lipman attempted these studies is that he was very interested in the possibility of life arriving on Earth from elsewhere in the galaxy, entombed in extraterrestrial rocks. The problems he faced are obvious. How can you be sure that the bacteria are not more recent contamination? And also, in the 1920s, the accurate dating of rocks was still in its infancy. If you plot the supposed age of revived rockbound bacteria against the year the study was done, you get an interesting pattern which illustrates these difficulties (Figure 76). All the oldest ages are from studies published in the first half of the twentieth century. Presumably, as methods improved, results from contamination by more recent

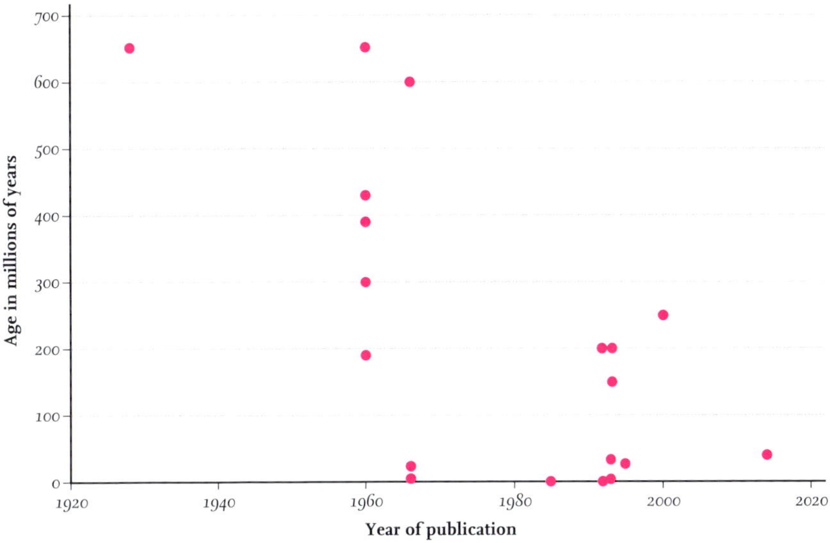

FIG 76. Claimed age of dormant microbes cultured from rock samples plotted against year of publication for studies claiming ages of at least 1 million years. Note how all the most extreme ages are from the earlier studies, when the available technology for such work was less advanced, suggesting they may be overestimates. (Redrawn from Wilkinson, 2023)

bacteria became less likely to make it to publication. After the 1960s, ages of a few tens to a couple of hundred million years were still being published, but not ones suggesting 300 million or more years. Many of the more recent dates of tens of millions of years are still controversial, because of concerns over contamination, despite the great methodological improvements since Lipman's time. Studies showing survival for hundreds to thousands of years (not shown on the graph), are much commoner and usually less controversial. Survival for a few million years seems very likely, and longer-term survival certainly seems possible for these 'undead' bacteria.

KOKKO'S QUESTION: THINKING ABOUT LIFE IN ALL ITS DIVERSITY

In 2017 Hanna Kokko published a research paper with the provocative title of 'Give one species the task to come up with a theory that spans them all: what good can come of that?'[28] Her point was that much of evolutionary theory was based on animals, often vertebrates, which are a tiny fraction of global biodiversity; so that the inspiration for much of our understanding of evolution is heavily biased towards rather untypical organisms that just happen to look rather like us. The discussion of senescence and sex in this chapter illustrates her point – the key theorists mentioned such as J. B. S. Haldane, Peter Medawar, John Maynard Smith and Bill Hamilton were all primarily zoologists. One aim of this book is to try and address this problem by making the natural history of microbes better known, and the discussion of sex, death and senescence shows why our vertebrate-based view of things may be very atypical of life in general.

Away from larger animals, senescence can be harder to find for many organisms, even when we keep them in captivity to reduce mortality from predators and parasites. However, as Graham Bell showed by assembling results from many early-twentieth-century studies, asexual clones of protozoa kept in captivity did seem to show declines in their ability to reproduce over time. He suggested that sex was often crucial in the long-term survival of protist populations. The distribution of both senescence and sex in the microworld is still poorly known, but a reasonable summary of research in recent decades is that the more we look, the more examples we find. Certainly the variety of sex in the microworld adds to our understanding of biology in general. Even many biologists, educated on a highly biased vertebrate-rich curriculum, unthinkingly link sex with reproduction. However, for much of biology the two are not closely linked. In many protists much of the reproduction is asexual, with often occasional sex presumably to try and counteract problems such as Muller's ratchet. The multiple mating types in ciliates and slime moulds also undermine the idea that there are only ever two sexes in a sexually reproducing species.

When it comes to death, the widespread occurrence of apparently reanimated zombies in the microworld raises fascinating questions about the definitions of life and death. However, the ability of many microbes (including some microscopic animals) to sit out hostile conditions in apparently metabolically inactive states, and then come back to life when conditions improve, is a key aspect of their natural history.

The microworld is also important for understanding the evolution of large multicellular organisms, as all multicellular individuals have single-celled microbes as ancestors. This forms the subject of the next chapter.

FIG 78. Discoidal Ediacaran fossil in Bradgate Park, Leicestershire. Such disc-shaped fossils are widely found in Ediacaran sites around the world, but their interpretation has proved difficult. Before more obvious fossils were found in these rocks, many people wondered if they were fossils at all. They were identified as jellyfish in the mid-twentieth century, an interpretation now widely believed to be incorrect, and it's now thought that they comprise a range of different types. Some of the fossils appear to be attachment discs ('holdfasts') which have broken off larger organisms, while others may be microbial colonies, fungi or early relatives of some marine animal groups (MacGabhann, 2007).

a schoolboy, accompanying my father, who had an amateur interest in geology, on a Manchester Geological Association field trip.

The discovery of *Charnia* had implications for the interpretation of some strange Australian fossils. A young geologist called Reg Sprigg had found similar-looking fossils in the Ediacara Hills in the Flinders Ranges in South Australia in the 1940s. He realised they were old, but somewhat cautiously suggested they were likely to be early Cambrian – the time period when the first visible fossils were thought to turn up. The publication of the Charnwood fossil with a clearly Precambrian date caused these Australian fossils to be re-evaluated. Once it became clear that they were also Precambrian, they began to attract much more detailed study. Fossils of this type came to be called the Ediacaran biota – which is now known from some 40 sites around the world.

The fossils from the Ediacara Hills were initially interpreted as early examples of various animal groups, such as jellyfish, soft corals and worms. However, in the early 1980s Dolf Seilacher strongly argued that the Ediacarans were something entirely different, and not closely related to modern animal groups. He suggested that they were effectively an early evolutionary experiment in microbes evolving into larger organisms, which had left no modern descendants. This view was, at the time, enthusiastically accepted and popularised by the influential palaeontologist Stephen Jay Gould, who described sitting 'spellbound' listening to a conference talk by Seilacher as 'wave after wave of expanded meaning cascaded over me'.[4] Seilacher suggested that the Ediacaran fossils were something so different that they should be classified as an extinct phylum, which he called the Vendobionta; others have suggested a wide range of potential affiliations, ranging from lichens to giant protists. Currently the consensus is that the Ediacaran biota contains a range of types, some that probably have no modern descendants and others that are precursors to some modern groups – at least some of which are animals. However, the fossils are enigmatic, and the absence of convincing internal anatomy has made their interpretation particularly difficult.[5]

These Ediacarans provide a tantalising, if confusing, glimpse of the origin of large multicellular organisms. This is arguably one of the key steps in the evolution of life on Earth. In a posthumously published essay, at the end of a very long career mainly working on microscopic protists, the biologist John Tyler Bonner suggested that there were four major 'eras' in the history of life on Earth.[6] Two of these have already featured in chapters in this book – the first life and the evolution of eukaryotic sex. His third big step was the evolution of multicellularity (his final step, the evolution of animal nervous systems, is not relevant to this book). However, we don't need to rely on fossils alone to start to understand multicellularity, for our current microworld contains many examples of cells gathering together into simple multicellular structures. Hunting around in the damp places at Bradgate Park, or many other locations, will reveal examples. Most spectacular are some of the slime moulds, with their fungal-like fruiting bodies, but there are algal examples too. It is these which I will consider first.

GILBERT WHITE AND THE WITCHES' BUTTER

In damp conditions the cyanobacterium *Nostoc* can form visible gelatinous structures. With many cells gathered together to make an organism that needs no magnification to be visible to the naked eye, it creates a blue-green slimy covering on paths and vegetation (Figure 79). Put some under a microscope and it resolves into a collection of cyanobacterial cells, but without magnification it's not obvious what you are looking at. Gilbert White – the eighteenth-century cleric

FIG 79. *Nostoc*, growing in damp conditions on a concrete surface in my Nottinghamshire garden. These sheets are often composed of *Nostoc commune* – which is the type species for the genus *Nostoc* (i.e. the species used to define the genus). Over the last 10 years or so this traditionally large genus has been the subject of many studies by molecular biologists interested in cyanobacterial classification – and as a result this well-known classical genus is now split into a confusing 16 or more genera (Kaštovský, 2024). However, as it is the type species, *N. commune* by definition stays in the genus *Nostoc* (see also Figure 80).

and author of *The Natural History of Selborne* – repeatedly recorded it in his journal as growing in damp conditions on grass and paths around his home in Selborne, Hampshire. While he did not include *Nostoc* in the letters that comprised his famous book, he did discuss it in his correspondence with various friends. He clearly thought it some sort of 'plant', probably fungal, as he refers to it as *Tremella nostoc* (*Tremella* is a genus that contains several gelatinous fungi). Indeed, if I didn't already know what it was, and didn't have access to a microscope, then some sort of gelatinous fungus or lichen would be a reasonable guess.

By the time White was writing, many naturalists were starting to view *Nostoc* as some sort of alga (White seldom had access to a microscope, just occasionally using one belonging to the Curtis family in nearby Alton, but to look at things such as sand grains rather than the microworld of this book). These *Nostoc* jellies were striking enough to have attracted various common names such as 'witches' butter'[7] or 'star slutch', the latter suggesting it was the remains of fallen shooting

FIG 80. Spherical *Nostoc* colonies – *Nostoc* pearls – in a small pool at Gaitbarrows nature reserve on the southern Cumbrian limestone. In (a) the oak leaf in the foreground in the pool gives an indication of scale, while (b) shows the *Nostoc* pearls in close-up – also nicely illustrating why such cyanobacteria used to be called blue-green algae. Such colonies are quite common in pools in limestone, and are often comprised of *Nostoc commune*. This, and related species such as *N. flagelliforme*, are eaten in Chinese and other Asian cuisines (Whitton & Potts, 2000). (Alex Hyde)

stars. White related that 'apparently intelligent' people considered it to be 'a mass of undigested food caste-up out of the stomach of cows', a once common idea that was dying out by the eighteenth century. One of White's friends had suggested to him that witches' butter was somehow produced by frogs.[8] Clearly its exact nature was still rather uncertain to many people at the time.

As well as the gelatinous sheets seen by Gilbert White around eighteenth-century Selborne, in freshwater pools *Nostoc* can sometimes form pea-like colonies often called *Nostoc* pearls or grapes (Figure 80). These sheets and pearls of *Nostoc* are visible without a microscope because a large number of cells are grouped together to form a single colony. This is the absolute minimum you need for something that you might describe as multicellular. However, some cyanobacteria take things a step further with different cells performing different functions. And once you have such a division of labour then most biologists would be happy to describe the organism as properly multicellular.[9] For example, cyanobacteria can sometimes form specialised cells called heterocysts in which nitrogen fixation takes place (see Figure 65 – the heterocysts are the larger circular cells in the algal filaments). The chemistry of fixing nitrogen – taking nitrogen from the air and turning it into a chemical form that a cell can use – requires an absence of oxygen. As cyanobacteria give off oxygen as a by-product of photosynthesis this is potentially a problem, and one solution is to fix nitrogen in specialist cells from which oxygen can be excluded. This ability of some cyanobacteria to fix nitrogen is important in the ecology of cyanobacterial blooms, discussed in the next chapter.

The interesting thing about cyanobacteria for the current chapter is that they are examples of prokaryotes on the edge of multicellularity – with cells grouped together into chains or sheets, and sometimes a simple division of labour between cells. However, the larger multicellular organisms we are familiar with, such as plants and animals, are eukaryotes, not prokaryotes. A classic, informative, and very beautiful microscopic eukaryotic alga that is similarly on the edge of multicellularity is *Volvox*.

ADAM SMITH AND THE FIERCE ROLLERS

The green alga *Volvox* has been a microscopist's favourite ever since it rolled into view at the very start of microscopy (Figure 81). As the Reverend W. W. Spicer wrote over 150 years ago, 'Their singular motions ... their peculiar forms, and wonderful modes of propagation, render them objects of great interest to the microscopist' (Figure 82).[10] It was first described by van Leeuwenhoek at the start of the eighteenth century, and its name combines the Latin *volvere* ('to roll')

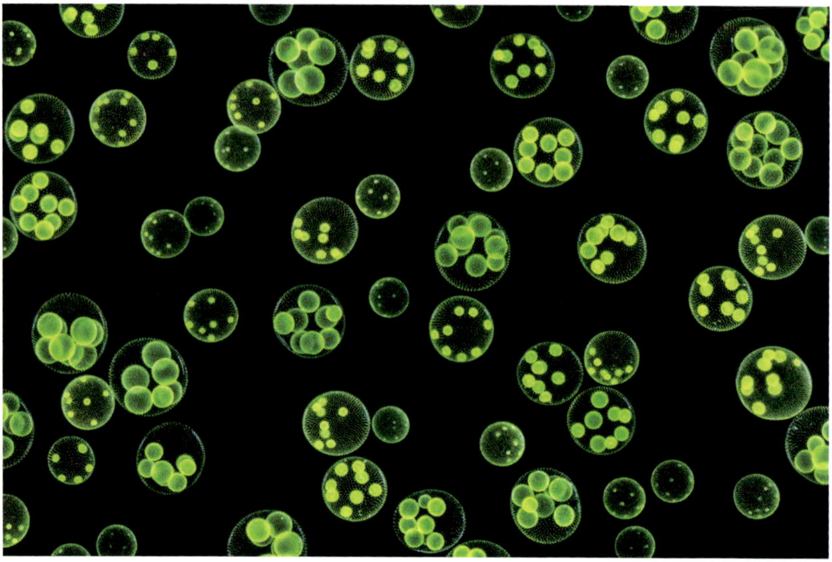

FIG 81. *Volvox aureus.* The smaller spheres within the larger ones are so-called 'daughter colonies': these are the product of asexual reproduction and are released when the larger sphere breaks apart. When triggered by environmental conditions the gonidia which give rise to the asexually produced daughter colonies can instead form either a sexual female (egg-containing) or male (sperm-containing) individual (Umen & Herron, 2021). (Alex Hyde)

with *–ox*, from *ferox* or *atrox* ('fierce'). Fierce rollers is a good description of its motion, as it's a spherical multicellular alga that swims with a distinctive rolling motion. It has become an important experimental organism for people interested in understanding the evolution of multicellular organisms.[11] The largest colonies can be up to 2.5 mm in diameter, and can be composed of thousands of individual cells. It's a common planktonic genus most often found in still water; small ponds and ditches are good places to look for it. During colder months it often overwinters at the bottom of a pond or lake as a zygote – that is, a cell produced by sexual reproduction.[12]

The ability to move ('roll') in *Volvox* species is important to their natural history. As photosynthetic green algae they need to access both light and nutrients, which they get by moving through the waterbody in search of the best conditions. They move with the aid of whip-like flagella – each cell having a pair of these. Here is one potential advantage of being multicellular: the *Volvox* colony can move faster than the individual cells would be able to do because there is an approximate positive relationship between speed of movement and size of organism.[13]

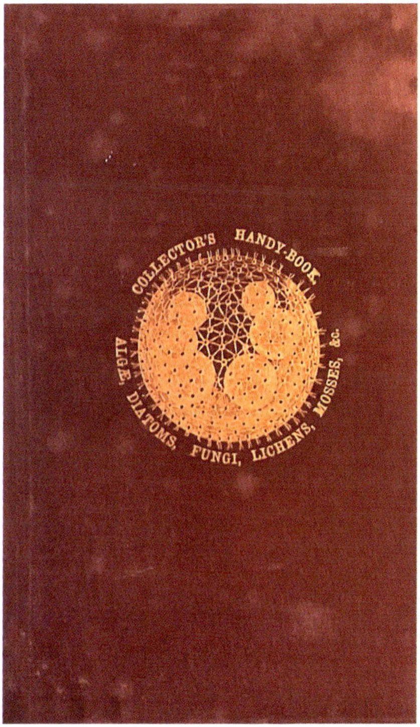

FIG 82. *Volvox* embossed in gold on the cover of Johann Nave's *Handy-Book to the Collection and Preparation of Freshwater and Marine Algae, Diatoms, Desmids, Fungi, Lichens and Mosses* (English translation, 1867). It was very much a practical book, detailing how to find and preserve your specimens – hence the water staining on the cover of the copy shown here. Nave was an expert on algae and a friend of Gregor Mendel (they met as students in botany lectures at the University of Vienna). Nave wrote the book in German while he was housebound, dying of tuberculosis, and it was translated into English by the Reverend W. W. Spicer (van Dijk & Ellis, 2020). *Volvox* didn't feature in Nave's original manuscript, but Spicer was so taken with the alga that he inserted an editorial note describing it and decorated the cover of the English edition with its image.

Therefore *Volvox* colonies are able to move to track changes in light and nutrients more effectively than closely related single-celled algae such as *Chlamydomonas*. It's possible that this speed of movement may also help them escape predators; certainly their size makes them too large for some potential predators to handle. This idea was illustrated by a well-known experiment from the 1990s in which laboratory cultures of the green alga *Chlorella vulgaris* were subjected to predation by flagellate protozoa. In less than 100 generations of the algae, instead of a single-celled growth form, a multicellular form took over which was too large for the protozoan to attack.[14] So evolving ways to stick your cells together to make a multicellular mass may sometimes be an anti-predator adaptation.

Moving more quickly and reducing predation are not the only potential advantages of becoming multicellular. *Volvox* has two different types of cell; most are photosynthetic cells with flagella, which form the sphere, but it also contains reproductive cells (gonidia) which give rise to the 'daughter colonies' that can be seen in Figure 81. A common way to understand the potential benefits of having specialised cells is by drawing analogies with ideas on the division of labour in human society associated with the eighteenth-century economist Adam Smith.[15] The idea was that productivity could be greatly increased if tasks that were all performed by a single person were split between several people, each of whom carried out a limited number of simple repetitive tasks. The canonical example is making a pin – instead of one person making the whole thing, different parts of the process are carried out by several people, leading to a streamlining of the process and a significant increase in pin production. It was also suggested that a market – for pins or whatever – would need to be of a certain size before the benefits of such a division of labour would become apparent. How might this apply in an alga? In *Chlamydomonas* the single cell has a temporary division of labour, as it can either be photosynthetic and motile or switch to being a reproductive cell. However, *Volvox*, with its larger size and two types of cell, can conduct both functions at the same time. This has the potential to lead to greater reproductive success.

So *Volvox* can provide some clues relevant to the evolution of multicellular organisms, and we don't have to rely solely on enigmatic Precambrian fossils in our attempts to understand the issues. *Volvox* is not unique in this respect, as multicellularity has evolved multiple times among the green algae – and it may indeed have evolved several times within the current genus *Volvox*. Another example already discussed can be seen in the stoneworts described in Chapter 4 (see Figures 38 and 39). Here, algae have taken on a form that looks very similar to vascular plants. Green seaweeds provide another more familiar example of multicellular forms of green algae.

Do these various algae allow us to say anything more general about how multicellular organisms evolve? It seems that what the multiple green algal examples tell us is that there are many ways in which a single-celled photosynthetic organism can acquire multicellularity – leading to organisms as different as *Volvox* and the stonewort *Nitella*. As one recent review of the topic put it, 'If there is one universal principle that emerges from comparing the evolution of multicellularity across green algal taxa, it is that there are no universal principles.'[16] However, algae are not the only microbes where simple multicellularity can be found. Some of the slime moulds are classic protozoan examples, and I will consider them next, before returning to more general points at the end of the chapter.

SLIME MOULDS: THE SOCIAL AMOEBAE

Slime moulds (or slime molds in American English) are formed by amoebae which gather together to produce very fungal-like fruiting bodies. Unusually for microbes, some of them form structures that are easily large enough to be visible to the naked eye. The exact nature of these organisms proved something of a problem, and historically they were often described as fungi, at a time when fungi themselves were considered to be plants (Figure 83). However, it's now clear that the main types of slime moulds are true amoebae (in the Amoebozoa – see Figure 10) and fall into three main groups. These are the protostelid slime moulds, the cellular slime moulds (the dictyostelids) and the plasmodial slime moulds (the myxogastids or myxomycetes); in addition, there are several other slime-mould-like social amoebae spread across multiple eukaryotic supergroups.[17] The simplest and smallest amoebozoan slime moulds are the protostelids, which were only discovered in the mid-twentieth century, and are relatively poorly studied. Unlike the other two larger types they have small cells which give rise to a small fruiting body which produces only a single spore. However, it is the cellular and plasmodial slime moulds that are most relevant to this chapter's theme, as during their life cycle they form larger more complex structures. They are also ecologically significant, as both groups are thought to be important predators of bacteria in soils.[18]

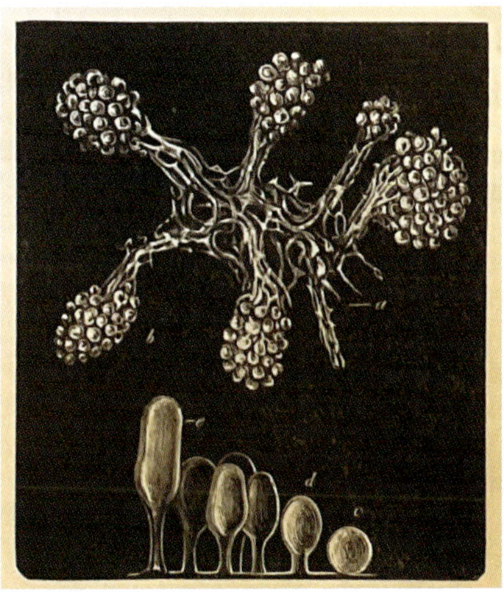

FIG 83. Slime moulds – in this case a plasmodial slime mould in the genus *Stemonitis* (from Warming, 1890, p. 58). Warming refers to these as a type of fungus, although he did note their resemblance to amoebae (then considered animals), which gave them a somewhat ambiguous position – so they had 'quite an isolated position in the vegetable kingdom'– part way between plant and animal. Because of this history, many slime mould papers are still often published in mycology journals, despite the fact they are no longer considered fungi. The figure shows the plasmodium at the top, with the fruiting bodies below. Warming's *Handbook of Systematic Botany* was an influential text on plant diversity and classification at the end of the nineteenth century. First published in Danish, it achieved much wider influence when translated into German (arguably the major scientific language at the time; many British botanists read German, and would have used this translation). This figure has been reproduced from a copy of the book that originally belonged to the influential botanist and ecologist Arthur Tansley. The handbook was translated into English five years later in 1895.

The plasmodial slime moulds are the ones that produce the large obvious bodies that can be seen in the wild without the need for even a hand lens. These will be described in the next section. Although the cellular slime moulds are smaller – they often have fruiting bodies that require at least a hand lens to see any detail, and are close to impossible to find in the field – they are fascinating, and have become key experimental organisms for researchers interested in evolution and developmental biology. One particular species, *Dictyostelium discoideum*, has become the experimentalist's favourite. As an indication of its scientific popularity, I carried out a quick web search and found over 200 research papers published in 2024 with this species mentioned in the title, and many more which contained references to it within the article itself. Because of this research attention, accounts of cellular slime moulds often tend to become a biography of *D. discoideum* – however, there are over 200 other species known (likely an underestimate), many of which are very poorly studied.[19] Some idea of how our knowledge of their diversity has grown over the last century or so comes from John Tyler Bonner's 2009 book *The Social Amoebae*. In it, Bonner (who was a leading cellular slime mould researcher) points out that around 1900 there 'were about seven species' known, and by the early 1940s, when he started to study them as a young scientist, not much had changed and there were 'about ten' species.[20] However, by the time he was writing the book in his late seventies, 'about 100 had been identified' – and in less than two decades this has doubled to the current figure of around 200.

It's their life cycle that has made cellular slime moulds in general, and *D. discoideum* in particular, of such great interest. They spend part of their life as individual single-celled amoebae living in the soil, but also form simple multicellular organisms to create their fruiting bodies and disperse their spores. These spores are sticky, and are often dispersed by animals, either on their surface or internally if eaten.[21] The life cycle of the well-studied *D. discoideum* is summarised in Figure 84. Spores give rise to free-living amoebae which feed on bacteria and other very small soil organisms. When food supplies are starting to run short the free-living amoebae aggregate and form a multicellular 'slug'. This can be anything from around 0.2 mm to 2 mm in length and can typically consist of between 10,000 and 2 million cells. The slug is attracted towards light, causing it to emerge from the soil and leaf litter, at which point it can develop into a fruiting body. This comprises a stalk with spore-forming cells at its tip. In the case of *D. discoideum* there is also a disc-like base to the stalk – which gives this species its specific name *discoideum*. The fruiting body releases spores, and the cycle starts again.

The life cycle of *D. discoideum* has particularly attracted the attention of biologists interested in studying the development of multicellular organisms, as a collection of free-living amoebae aggregate to form a multicellular organism

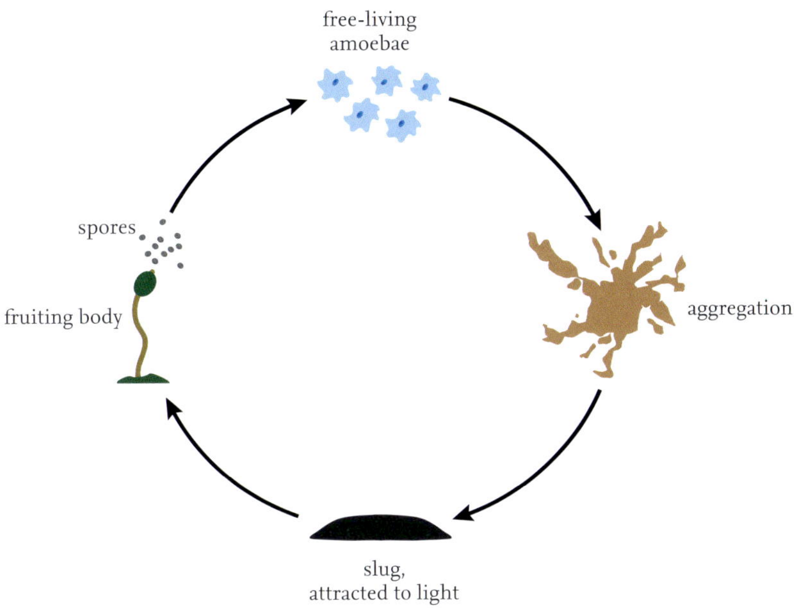

FIG 84. The social cycle of *Dictyostelium discoideum*. The diagram, and the description in the text, are based on several sources, especially Romeralo *et al.* (2012) and Bonner (1969, 2009). See text for more details, and note that the different stages are not drawn to scale. Much of the reproduction is asexual, as shown here, but sexual reproduction also takes place. Diagrams such as this are common in biology textbooks but give a somewhat misleading impression of the species' natural history. As well as cells that form 'slugs' and fruiting bodies, there are also cells that stay as single-celled amoebae throughout their lives, reproducing by dividing in typical amoeba fashion (Tarnita *et al.*, 2015).

and then the cells differentiate into different types, such as stalk- or spore-forming cells. This can provide a simple model for the much more complex developmental processes in an animal such as a human. However, the focus of this book is protist natural history rather than developmental biology.

For an organism that has been so well studied in the lab (for example, its full genetic makeup has been known for over 20 years), we know surprisingly little about its natural history in the wild. It was first found in North Carolina in the 1930s, and has a rather curious global distribution which also illustrates how our knowledge has changed over the last few decades. At the end of the twentieth century a global synthesis of the distribution of cellular slime moulds in forest soils showed *D. discoideum* with a distribution split between eastern

North America and east Asia (Japan and southern China).[22] Since then, its known range has increased but the split (technically called disjunct) distribution is still apparent – the species is now known to occur more widely across the Americas, and in Asia it's now found in India, Pakistan and Nepal, as well as China and Japan.[23] But it is still apparently absent from Europe, despite this region being the subject of several recent studies, and is also missing from other parts of the world. It's hard to know what to make of such a disjunct distribution. In larger organisms such a pattern is often taken to be an indication of a declining range size – where the organism was more extensively distributed in the past. However, as discussed in Chapter 5, both a shortage of data and the problem of cryptic species make interpreting large-scale patterns in protist distribution really difficult. So it is possible that this apparent disjunct distribution is at least in part an artefact of our limited knowledge.

Molecular methods have had an important effect on surveying cellular slime moulds in soils and leaf litter. I am not aware of any studies of this type in Britain, but in the Iberian Peninsula the diversity of cellular slime moulds was found to be highest in colder, wetter environments.[24] There appears to be a decline in diversity in Europe as one moves north, with a study in northern Sweden finding fewer species than identified in southern Europe. Nineteen species were recorded from the Iberian Peninsula, while the Swedish study found nine species, all but three of which were new to Sweden, and two of which were previously undescribed and new to science.[25] This gives some indication of how understudied this group is, even in Europe. In many parts of the world there are even fewer studies, though currently China appears to be one of the more active countries for research on the taxonomy and diversity of cellular slime moulds.[26]

As already described, things are very different in the lab, where many studies have been undertaken, especially of *D. discoideum* – and some of these are highly relevant to this chapter's theme of the evolution of multicellularity. For many decades it was clear the cellular slime moulds' life cycle raised an interesting evolutionary question. The problem is that many different amoebae combine together, apparently cooperatively, to make the fruiting body. This may not be a problem to those that end up producing spores, but those that form the stalk are helping other cells to reproduce while not reproducing themselves. That seems a bit odd from the point of view of natural selection. One key strand of my research during the 1990s was theoretical work on the evolution of cooperation, and a colleague and I realised that *D. discoideum* would make a very promising experimental subject for investigating some of these ideas. We had just started on an initial draft of a grant proposal to try and fund such experiments when Joan Strassmann and colleagues published a research paper in *Nature* describing the results of experiments very similar to what we had in mind.[27] It's very common in science for multiple people to hit on effectively the same idea around the same time.

One way to make the problem partly go away would be for all the amoebae that form a fruiting body to be genetically identical, and indeed this is often – though not always – the case. That's part of the solution in animals and plants, which are the multicellular organisms we are most familiar with. If all the cells have arisen from one egg and are therefore clones, their genetic interests should match. In such a case the stalk cells would be helping genetically identical kin to reproduce, which from an evolutionary point of view can be little different from reproducing yourself. But Strassmann and colleagues showed that not all slime mould slugs and fruiting bodies are clones, and they also experimentally created ones that had a mixed genetic background. They used molecular methods to follow the fate of these different amoebae and found that there was competition in who became stalk and who produces spores – so things were not as cooperative as might appear at first sight. This raises another interesting problem. If a small number of competitive strains win out in forming the next generation, over time you would expect the genetic diversity of the species to decline – as less competitive strains become extinct. However, this does not appear to be the case in the wild, so there are presumably other factors at work counteracting this trend.

Corina Tarnita and colleagues have suggested some potential solutions.[28] Contrary to the impression you might get from some biology textbooks, not all *D. discoideum* amoebae form slugs and fruiting bodies when food is short. Some persist as free-living amoebae and can reproduce by cell division. If food is slow to replenish following scarcity, then spore formation, with a chance of being dispersed to somewhere with more food, may be a good option. However, if the scarcity is short-lived then the amoebae that don't form fruiting bodies will quickly find food again, and may do better than the slug-forming amoebae. As bacterial populations can often grow quickly, this rapid replenishment may be reasonably common in the wild. Therefore the balance between these two life histories, slug-forming and non-slug-forming, may help maintain diversity – as the adaptations for competing well in a fruiting body may be harmful when free-living.

These are not the only interesting differences between different clones of *D. discoideum*. In 2011 a North American study, again involving Joan Strassmann and her husband David Queller, found that around one-third of the slime moulds collected from the wild were indulging in what the authors described as 'primitive agriculture'.[29] When their bacterial food was starting to run out these amoebae formed a slug early, while there was still some food left. They incorporated some of these bacteria into the slug, and when they formed a fruiting body bacteria could be found in their spores. This meant that the amoebae colonising a new site by spore dispersal took a starter pack of bacterial food with them – inoculating their new home with a population of edible bacteria. The suggestion is that the reason that all clones don't do this is that there are disadvantages as well as advantages to taking bacteria with you when you disperse. If there is no suitable food at the new site, then inoculating it with edible bacteria is clearly advantageous. However, to do this – that is, to have bacteria available – the clones have to stop feeding earlier, so they don't consume all available food before forming a slug. This cost is not worth it if their spores arrive at a site with plentiful food already there. So in some cases we would expect the 'agriculturalists' to do better, but in other cases they will do worse.

Cellular slime moulds clearly show a range of interesting multicellular adaptations at various points in their life cycle. However, the slime moulds that a naturalist is likely to come across in the field are plasmodial slime moulds – which have evolved a rather different approach to becoming large and complex.

DEMON DROPPINGS? THE PLASMODIAL SLIME MOULDS

The slime mould that is most often seen in Britain is probably *Fuligo septica* (Figure 85). The fact that it's both quite common, and often bright yellow, makes it easy to spot. Indeed it's prominent enough to have attracted multiple English names, most commonly 'dog-vomit slime mould', 'scrambled-eggs slime mould' or 'flowers of tan' – the latter name arising from its French name *fleur de tan*. This species is very widely distributed, and there is a long list of vernacular names from around the world – including 'fairy butter' and 'star slutch' (a name shared with *Nostoc*, as described above) from Britain, 'shit of witch' from Estonia, and 'ghost vomit' from Russia.[30] It seems likely it has more vernacular names than any other amoeba. The species was first described by science in France in the eighteenth century, although there are much older records of something that sounds very like *F. septica*. Several ancient Chinese medical texts describe 'kwei hi' (demon droppings) that are yellow in colour, and growing in shady damp places – the earliest known reference coming from the year 739 during the Tang Dynasty. It's widely suggested that these demon droppings are *F. septica*-like slime moulds.[31]

Plasmodial slime moulds have a very different approach to making a large 'microbe' which can give rise to the fruiting bodies needed for spore dispersal. Starting with a spore, a single-celled organism is produced which feeds on bacteria and other small soil organisms. This can switch between an amoeba-like form (in drier conditions), and a flagellated form when more soil water is available for 'swimming' in. When it's very dry these two forms can give rise to a resting stage (a cyst) until things become damper again. Under certain conditions, not well understood, a much larger body called a plasmodium is produced. But unlike the cellular slime moulds, where the slug is formed by many amoeba cells combining to form a multicellular structure (hence the term *cellular* slime mould), the plasmodium (e.g. the large yellow structure in Figure 85) is effectively one large cell containing multiple nuclei – which can move, slowly flowing over its substrate. The plasmodium can often be formed via sexual reproduction, or sometimes has its origin in asexual reproduction by a single amoeba. As it grows larger it comes to contain an ever-growing number of cell nuclei. Some of these plasmodia are among the largest known cells. For example *Brefeldia maxima* (a species found in Britain) can be as much as 1–2 metres across. Fruiting bodies

FIG 85. Plasmodium of *Fuligo septica* in woodlands at the Centre for Alternative Technology, near Machynlleth, north Wales. It can be found at most times of the year but is commonest in warm, damp summer conditions. The photograph shows a typical habitat for this species, described by Ing (2020) as 'rotten wood, sawdust, forest litter and mosses (above buried wood)'. In the past, piles of oak bark used in leather tanning would sometimes be covered in this species, one of its many vernacular names being 'flowers of tan'.

FIG 86. Fruiting bodies of *Trichia decipiens*, sometimes called the salmon slime mould, in woodland in the Scottish Highlands. This is a common slime mould found on dead wood, with a very wide distribution – although it is usually only found at altitude in the tropics (Ing, 2020). (Alex Hyde)

for spore dispersal can arise from the plasmodium, and are very important in the morphological identification of species. The spores are released from the fruiting bodies and the cycle starts again.[32]

Fuligo septica is typical of plasmodial slime moulds in liking a woodland habitat. Dead wood, leaf litter and tree bark provide the habitat for many British species (Figures 86 and 87). A lower diversity can be found in habitats such as grasslands and heathlands. There are also species characteristic of more specialist habitats. As with the algae described in Chapter 2 associated with long-lasting snowbeds, there are also plasmodial slime moulds found on British mountains where there are snow patches that last for at least a few months over winter. In 2003, Bruce Ing catalogued, with an identification key, 21 British snowbed species (compared to around 60 in the Alps).[33] Cairngorm was the best British site, with 16 species, although most Scottish mountains had at least a few species – but only two species are known from England, and none from Wales. As with the specialist snow algae, climate change is clearly a threat to these slime moulds; they seem to be increasingly rare in Britain and of real conservation concern.[34]

FIG 87. *Ceratiomyxa fruticulosa*, sometimes called the coral slime mould, on a decaying log in the Peak District. This is another very common slime mould with a wide global distribution (Ing, 2020). (Alex Hyde)

What about diversity in less extreme habitats? The area around Chester provides a good example of the diversity that can be found in more lowland habitats. This region has been especially well studied because for many years the slime mould specialist Bruce Ing was based at the University of Chester. Cheshire itself has 182 recorded species, while Shropshire has 174, and South Lancashire 145.[35] To give these numbers a context, the current number of species known for England is 340.[36] Plasmodial slime mould diversity, extinction and conservation are discussed in more detail in the final chapter.

In the context of this book these slime moulds are important because in their plasmodial stage they are of a size which brings them to the notice of field naturalists (see the Appendix for suggestions on how to grow them from bark and leaf-litter samples). Like lichens, they form a bridge between the micro- and macroworlds. In the context of this chapter they are interesting because they have a very unusual way of becoming large, while remaining single-celled. The bigger question, however, is this: what do the various examples described in this chapter – from *Nostoc*, to *Volvox*, to various types of slime moulds – tell us about the evolution of larger organisms of the kind that more usually feature in the New Naturalist series?

THE MANY PATHS TO MULTICELLULAR LIFE

The Ediacaran fossils from Charnwood Forest show that it's not straightforward to understand the evolution of multicellular organisms relying on the fossil record alone. However, the various multicellular microbes described in this chapter can help us in thinking about how multicellularity evolves. The simplest examples involve cells sticking together as they divide to form a larger organism, but alternatively cells can come together and aggregate – as in slug formation in cellular slime moulds. There are likely many advantages to be gained from being multicellular. For example, the larger mats of *Nostoc*, as seen in Figure 79, will dry out more slowly than a single cell would. In addition, there are potential anti-predator advantages, as already described for the green alga *Chlorella vulgaris* – an idea strikingly illustrated by the water fleas and *Volvox* in Figure 88. *Volvox* also provides a good example of the potential benefits of division of labour in multicellular organisms, where different cells can perform different functions at the same time. However, most of the organisms described in this book are not multicellular, and these make up most of global biodiversity, so being single-celled is often a very successful lifestyle. Indeed the cellular slime moulds show that you don't have to be *either* single-celled *or* multicellular, but that a species can swap between both ways of life during its life cycle. One key message that can be drawn from the range of organisms described in this chapter is that over evolutionary history, simple multicellularity has evolved on many different occasions. The plasmodial slime moulds are particularly intriguing as they have a different strategy for becoming large: rather than increasing in size through sticking multiple cells together, their plasmodia form one giant cell with multiple nuclei.

All the examples discussed in this chapter are rather simple multicellular organisms when compared to the more familiar plants, animals and larger fungi. Perhaps complex multicellular organisms can only evolve under rather specific conditions? One important condition has already been discussed in the context of cellular slime moulds, which is that cells that are genetically identical have shared interests. This makes cooperation more likely than competition. We all started from a single fertilised egg that then divided multiple times to produce the complex multicellular organism that is a human. This means that all your cells should be genetically identical – give or take any mutations that happened during the large number of cell divisions needed to form, and then maintain, your body. Even in this case there can be problems, although starting with genetic uniformity there can still be mutations that benefit the reproduction of particular cells at the cost of significant damage to the individual organism, which is what gives rise to cancers.

The obvious way to evolve a single organism composed of genetically identical cells is for it to start life as a single cell, which is the approach taken by the familiar plants, animals and fungi. This obligatory single-celled genetic bottleneck between generations is largely absent in the examples discussed in this chapter. In addition, there is another potential problem. Consider multiple cells stuck together, which can reproduce by fragmenting into smaller groups of cells but are not forced by their life cycle to repeatedly pass through a single-celled bottleneck. As each cell independently acquires mutations over time, they will start to differ from each other and increasingly may not be able to successfully interact with one another.

And yet such cell-to-cell interactions are crucial for a successful multicellular organism. This is sometimes called 'the coordination hypothesis', which was first

FIG 88. Multicellularity as a potential anti-predator adaptation. By forming spheres that can contain thousands of individual cells, *Volvox* becomes too large to be easily eaten by predators such as water fleas *Daphnia* sp. The greenish gut contents running the length of the *Daphnia* show that they have been feeding on smaller algae. (Alex Hyde)

explicitly suggested in an essay published a quarter of a century ago by Wolpert & Szathmáry.[37] Being relatively recent, it has not been studied in the same detail as the idea about the importance of genetic uniformity to reduce conflict of interest, which has been around for much longer. Recently, however, the cellular slime mould *D. discoideum* has been used to test this idea, and this research produced evidence for such a lack of coordination leading to reduced spore production.[38] The giant multinucleated cell produced by plasmodial slime moulds suggests that these species should be particularly susceptible to such coordination problems, as their many nuclei acquire mutations over time.

The examples described in this chapter show that there are many ways of making simple multicellular organisms (or giant single-celled organisms in the case of plasmodial slime moulds), but the ideas I have just been discussing suggest that it may be hard for these to evolve into more complex multicellular organisms *unless* they have a life cycle where once a generation they pass through a single-celled stage (an 'egg'). This enforces genetic uniformity on the organism at the start of its life, so avoiding many of the genetic issues, such as competition and coordination between cells, that potentially make it so difficult to be a successful complex multicellular organism. However, these simple multicellular 'microbes' do bring the microbial world into a size bracket accessible to a naturalist without a microscope – so allowing us to glimpse microbial diversity in the field. This is important, as microbes are crucial to the functioning of ecosystems, but the danger is that because they are usually out of sight, they are too often out of mind as well.

CHAPTER 8

Ecosystems and Cycles

Early winter, and an unusually early snowfall catches the last of the carotenoid-yellow leaves on the trees – highlighted by the low-angled November sunshine. All is quiet, except for the noise of a distant road, but under the snow microbial life is already starting to work on the recently fallen leaves, albeit slowly because of the low temperature. I am in Bunny Old Wood in southern Nottinghamshire; it's ancient woodland on a slope too steep to be worth clearing for agriculture, so the breakdown of leaf litter here is an annual event that has been recurring for centuries (Figure 89). Another New Naturalist author, the great woodland historian and plant ecologist Oliver Rackham, also visited this wood in November, back in 1977 – recording an interview for Radio Nottingham.[1] What struck Rackham about the wood were not the invisible microbial processes under his feet, but the dominant role played by Wych Elm *Ulmus glabra*.[2] These trees are still a prominent part of the woodland almost 50 years later, having escaped the worst of Dutch elm disease.

Although the microbial breakdown of leaf litter was invisible to both Rackham and me, under certain conditions it comes into view, made visible by cold winter weather. A few winters before my 2024 visit to Bunny Old Wood, and only a kilometre or so to the north, on a freezing morning I witnessed microbial decay in action in the dung heap of a local farm (Figure 90). The steam rising from the dung was due to heat generated by microbes breaking down the organic matter that constitutes most of the dung heap. What I was seeing was the metabolism of colossal numbers of microbes, made visible in the cold winter air. Much of this microbial activity will be carried out by prokaryotes and fungi; these are the textbook microbes when it comes to the breakdown and recycling of organic material. However, as this chapter will make clear, recent research is

FIG 89. The Nottinghamshire Wildlife Trust's reserve at Bunny Old Wood in southern Nottinghamshire, after an early-winter snowfall. Deciduous woodland, such as this one, nicely illustrates the idea of nutrient cycling. Nutrients in the soil are used by, and incorporated into, the growing plants. The elements required in the largest quantity from the soil by plants are nitrogen (N), phosphorus (P) and potassium (K) – that is why so-called NPK fertilisers are important in gardening and agriculture (other elements are needed in smaller amounts – such as silicon (Si), which is discussed later in this chapter). Without processes to recycle chemicals in dead organic matter, crucial nutrients would eventually become unavailable, as they would be locked up in dead leaves, wood and other debris. Microbes play key roles in such nutrient cycling.

showing that protozoa and algae are more important in such processes than used to be thought.

On occasion the heat generated by the microbial breakdown of manure can be put to good use. Gilbert White, the eighteenth-century author of the classic *Natural History of Selborne*, was also a serious gardener, and managed to grow tender crops such as melons and cucumbers in his Hampshire garden – in a climate we now describe as the Little Ice Age. He did this using 'hot-beds', large beds filled with fresh manure which gives off heat as the manure is broken down by microbes, with soil placed on top into which the crop is planted (Figure 91). This was a large-scale operation: White's hot-beds were on occasion 14 metres long, and such large beds could require close to 50 cartloads of fresh farmyard manure. The unexpected thing is that in his records White describes this manure

FIG 90. A steaming winter dung heap on a farm in Nottinghamshire. The heat generated by microbial activity is made visible as steam in the cold winter air.

FIG 91. Small reconstructed hot-bed at The Wakes, Gilbert White's house, which is now a museum. These could be covered over to keep the heat in, and then opened to let out excess heat and let in the light. White often used a covering of translucent oiled paper, which helped keep in the heat but also allowed light to get to the plants (glass would have been very expensive to use with large hot-beds).

as 'borrowed' rather than bought. It seems he was paying for the delivery of the manure and its temporary use – effectively he was renting it, before returning it to the farmer once the hot-beds were finished with for that year.[3]

The state of the manure matters, depending on how you are planning to use it. This was known to Theophrastus 2,300 years ago, who wrote that 'the most pungent manure' is often bad for plants.[4] Modern gardeners know this well, and will likely be able to guess what was going on in eighteenth-century Selborne. Fresh manure is of limited use for fertilising plants; the manure you buy to put on your garden or allotment is often marketed as 'well-rotted manure'. This is manure that has already been subject to some microbial breakdown. So rather than the heat going to waste (as in Figure 90), White was using the heat to grow his melons and cucumbers, then returning much of the manure to the farmer in a state now fit for using to fertilise crops. A sustainable approach to food production that was to be largely replaced during the twentieth century by heating relying on fossil fuels, and fertilisers from the chemical industry.

A VIEW FROM THE OCEAN: THE MICROBIAL LOOP

This book focuses on the protists of soils and freshwater, but a key idea relevant to this chapter started in the sea. As the open ocean is unfamiliar to most naturalists, except on TV documentaries, instead visualise a large rock pool. If you drew a food chain for such a pool, what would it look like? The basic food chain we are introduced to in school science classes usually starts with a photosynthetic plant, which is then eaten by a herbivore, which is in turn eaten by one or more carnivores. So a simple deciduous woodland food chain, of the type you might find in Bunny Old Wood where this chapter opened, might look like:

Leaf → caterpillar → great tit → sparrowhawk

By analogy, a food chain for a large rock pool might go something like:

Alga → small crustacean → starfish → lobster

Up to around 1980 many ecologists would have agreed with this, although some marine biologists already realised that there were important microbial complications which needed including. Figure 92 is a visual representation of this hypothetical rock-pool food chain – although the mid-twentieth-century

FIG 92. Common Lobster *Homarus gammarus* in shallow water at Rhosneigr, Anglesey (note that the lobster has lost an antenna, presumably from a predator attack – so although it's the top predator in our hypothetical food chain it is also preyed on by other animals, including humans). The photograph shows both algae (the tufts of seaweed) and the lobster, which in the simple food chain feeds on starfish, which feed on smaller crustaceans, which in turn feed on algae – although as described in the text these are more likely to be microscopic algae rather than seaweeds.

view was that the most important algae were often larger microscopic protists such as diatoms, rather than the multicellular seaweeds you see when you look into such a pool. However, we now know there are some important additional complications.

A key aspect of these complications is the very high number of bacteria found in the ocean. A cubic metre of seawater can contain something like 100,000,000,000 bacterial cells (more conveniently written as 10^{11}). This is the same as the number of stars shining in a typical galaxy such as our own.[5] Western science was rather slow to realise the importance of marine bacteria, in part because the preferred experimental approach was to grow bacteria from water samples in culture on agar plates. In the East – especially in the Soviet Union – scientists were more likely to use direct counts of bacteria using a microscope, and so had realised by the 1960s that there were quite literally astronomically high numbers of marine bacteria. However, for both linguistic and political reasons, Western scientists were slow to pick up on this work.[6]

Along with the view that bacteria were not particularly important in the oceans, the received wisdom around 1970 was that protozoa were also unlikely to be important. The textbook marine food chain around this time was similar to the one shown on page 170, with small animals (e.g. crustaceans) feeding on algae such as diatoms, and these small animals then being eaten by larger animals. But in about 1980 a variety of discoveries coalesced into an idea called the microbial loop. It was realised that much of the organic matter produced through photosynthesis by microscopic algae (both by larger algae such as diatoms, and by many smaller ones – including lots of very small cyanobacteria) was not eaten by organisms feeding directly on the algae, but leaked into the surrounding water as chemicals usually described as 'dissolved organic matter'. It was also realised that much marine photosynthesis was carried out by algae so small that they couldn't be successfully eaten by the much larger 'small' crustaceans, such as copepods. However, the dissolved organic matter can be utilised by bacteria, which may then be eaten by smaller protozoa – such as smaller flagellates and some ciliates. In addition, these smaller protozoa can also feed on the smaller algae. The protozoa feeding on the bacteria and the smallest algae are then fed on by larger organisms such as near-microscopic crustaceans. So in addition to the simple food chain shown above there is also a loop that goes from algal photosynthesis (via dissolved organic matter) to bacteria which are then fed on by protozoa which provide food for tiny animals, at which point the nutrients the protozoa contain are fed back into the conventional food chain. At its simplest, the rock-pool food chain shown above, with this loop added, would look something like this:

These various ideas were drawn together in 1983 in a famous article entitled 'The ecological role of water-column microbes in the sea'.[7] Reflecting on the influence of this article a quarter of a century after its publication, one of its co-authors, Tom Fenchel (Figure 93), wrote that 'probably all the authors of the paper have since wondered why it was so successful'. He very plausibly suggested that the key reason was that they gave this emerging idea a name – the microbial loop. As he wrote:

> *The term 'microbial loop' suddenly clarified what we and others were working with and it serves as a label for studies of marine and freshwater microbial environmental microbiology.*[8]

FIG 93. Tom Fenchel, photographed in 2014, one of the authors of the paper that named the microbial loop (Azam *et al.*, 1983). As described in the main text, in a paper published 25 years later he suggested that the most important aspect of this work was that it gave the idea a name – the microbial loop – so helping to give research in this area a focus. See also Esteban & Fenchel (2020, pp. 107–108) for additional background.

Stephen Jay Gould wrote in a different context that, 'phenomena without names ... will probably not be recognised at all'.[9]

As the quotation from Tom Fenchel implies, these ideas apply to freshwaters as well as the sea – although some of the details may differ depending on the dominant type of zooplankton. But what about soils? The roots of plants leak large amounts of carbon compounds into the soil. As with the dissolved organic carbon compounds in seawater and freshwater, these are also fed on by bacteria, which are in turn fed on by protozoa. So there is also a microbial loop functioning in soils, which is especially pronounced around plant roots. In this system the protozoa are often playing an important role in making nutrients locked up in the bacteria available for use by plants. One type of protozoan that may be particulary important in affecting soil bacterial communities are the social amoebae (slime moulds), discussed in the previous chapter in the context of the evolution of multicellular organisms.[10] The important role played by protozoa in soil ecology, especially nutrient cycling, has become increasingly apparent in recent years.[11]

A VIEW FROM THE SOIL: PROTOZOA

The discussion of food chains, and the microbial loop, makes it clear that what protozoa eat is very important in understanding their role in ecosystems (Box 5). As already discussed in several chapters, when we think of a protozoan feeding, it is phagocytosis that most readily comes to mind. That is, the engulfing and uptake of food particles by the protozoan. In many amoebae phagocytosis can take place anywhere on the cell surface, while in ciliates it tends to happen at particular locations – informally referred to as the protozoan's 'mouth'. In addition, protozoa can absorb dissolved nutrients from their environment, and as discussed in Chapter 4 many can effectively photosynthesise via their symbiotic algae. At their most basic, protozoan feeding strategies can be split into three main types – raptorial feeding (active hunting), diffusion feeding, and filter feeding. In diffusion feeding the protozoan passively waits for food particles to make contact with it, while in filter feeding the often stationary protozoan creates water currents to bring food particles within reach.[12] Diffusion feeding is not common among the protozoa described in this book, although it is used by some ciliates. However, if this was a book on marine protists it would deserve much more attention, as it is used by a very important group of marine protozoa, the foraminifera.

Taking a 'view from the soil' requires a basic understanding of the nature of soils, and soil structure provides key components of the habitat in which the protists live. Two aspects are particularly important: namely, the distribution of pore sizes and the amount of water in the pores (Figure 95). One way to describe soil is as a collection of mineral particles – such as sand or clay – with gaps between them known as soil pores. These pores are of varying size and are filled with air and/or water. The microbes that feature in this book cover a wide range

Box 5. What is an ecosystem?

The key idea behind the concept of the ecosystem is that for many purposes living organisms are best understood as being part of a system where there is a two-way interaction between life and the physical and chemical components of the environment. For example, the acidity of a soil, or a waterbody, affects which organisms can survive at that location, but the organisms also affect acidity via chemicals they release into the environment. To understand the workings of the environment you need to understand the full system – not just one aspect, such as life or chemistry in isolation. Often, human actions have simplified the system (Figure 94).

FIG 94. Woodland and arable field ecosystems at Bunny Old Wood – the site where this chapter opened. The arable field is a much simpler system than the woodland. Indeed, in much of modern agriculture many of the ecosystem processes that would have been supplied by soil microbes and fungi in a less intensively modified system have been replaced by chemical fertilisers and pesticides.

As I wrote in a previous book in the New Naturalist series, I view ecosystems as 'one of the fundamental things – or entities – in ecology, comprising not only all of the organisms in that system but also its physics and chemistry and all of the interactions and feedbacks between these many components'.[13] Soil provides a nice example where 'organisms, inorganic particles, water and gases act together', and indeed in this respect the soil can be regarded as a model for the whole Earth system.[14] While ideas of feedbacks between plants and soil in agriculture have been written about for over 2,000 years, an understanding of their importance in ecosystems is more recent, forming 'an emerging, but little studied, area of research linking ecological and evolutionary processes'.[15]

The first use of the term *ecosystem* was in a 1935 paper by Arthur Tansley; as Cameron and Earley pointed out 80 years later, one of the interesting aspects of his idea was that he considered ecosystems to be both real (existing in nature), and a human construct (a figment of the scientists' imagination).[16] This still seems a realistic way of viewing ecosystems – as partly a simplification constructed for our own convenience when trying to describe a complex world, and partly describing important aspects of reality.[17]

of sizes – from bacteria, which may be only 1–2 μm wide or even less, to protozoa, which are usually some tens of micrometres wide, to small animals which prey on them, which may be several hundreds of micrometres across. While larger animals such as earthworms can push their way through the soil, the movement of smaller organisms is constrained by the availability of suitably sized pores through which they can move. Small pores can provide a refuge from predators. For example, a tiny pore only 1 μm or so across can contain small bacteria but not allow access by any of their predators. While animals the size of springtails can move through air-filled pores, the smaller organisms usually swim. So to move around, bacteria, protists and nematodes need either water-filled pores, or at least pores where the soil particles have a thin coating of water on their surface. This means that larger soil protozoa are especially affected by drought conditions; smaller types are able to continue to move and feed for longer in drought-affected soil, as they can make use of thinner water films coating the soil particles.[18]

How many microbes are living in these soil pores? Obviously their density in soil is very variable, changing with moisture and food availability among other factors; however, some ballpark figures will be useful in getting a feel for their importance in soil processes. The numbers of bacteria in a gram of soil can reach similar numbers to those cited above for a cubic metre of seawater – namely 10^{10} or 10^{11} individuals. At the soil surface, where densities are highest,

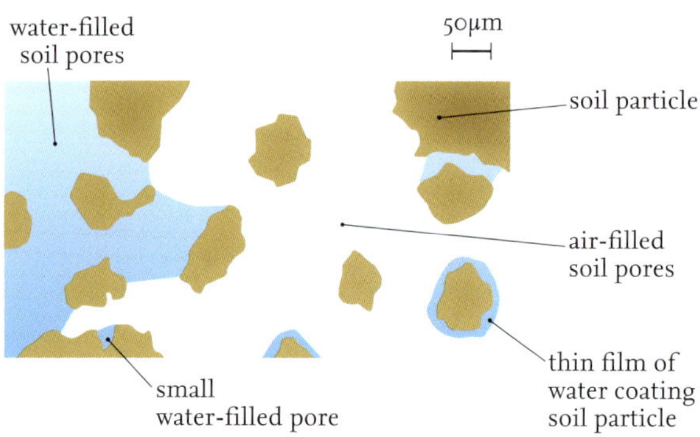

FIG 95. Organisms the size of springtails (typically 1–2 mm long or less) and smaller are limited by available soil pores (larger animals such as earthworms don't have this restriction, as they can force their way through the soil). Protozoa, along with some small animals such as nematodes and tardigrades, require water-filled pores, or at least soil particles with a thin film of water covering them. Animals such as mites and springtails are usually restricted to the air-filled soil pores. Small pores can provide refuge from larger predators – for example a pore with a diameter of less than 30 μm would usually provide protection to smaller protozoa from predatory nematodes. Based on multiple sources, especially Erktan *et al.* (2020) and Nielsen (2019).

numbers of protozoa can be between 10^4 and 10^6 per gram.[19] Because soils are dark, the assumption has been that algae are unlikely to be important in this environment – but, as described below, recent research is causing this apparently commonsense conclusion to be revised. However, before discussing these algae I will use testate amoebae to illustrate the importance of protozoa in soil nutrient cycling, as a general introduction to the importance of protozoa in soil processes.

TESTATE AMOEBAE AND SOIL NUTRIENT CYCLING

A good starting point in thinking about the importance of testate amoebae in soil processes is to ask what they eat. Traditionally testate amoebae were assumed to mainly eat bacteria; however, they have long been recorded eating a wide range of other foods, including other protozoa, algae and fungi. As can be seen from Table 3, some testates (such as *Nebela tincta*) appear to be generalists feeding on a wide range of food sources, while others (e.g. *Phryganella acropodia*) are much more specialist. There is also growing evidence that some feed on dead organic matter in soils, potentially playing an important role in the recycling of nutrients

TABLE 3. Food use by a selection of testate amoeba species. B, bacteria; Fu, fungi; Cb, cyanobacteria and other microalgae; Fl, flagellates; Ci, ciliates; T, testate amoebae; A, microscopic animals; O, organic matter; P, plant material.

Amoeba species	B	Fu	Cb	Fl	Ci	T	A	O	P
Centropyxis aerophila		X	X						
Corythion dubium	X			X					
Euglypha rotunda	X	X						X	
Hyalosphenia papilio *		X	X		X		X	X	
Nebela tincta		X	X	X	X	X	X	X	X
Phryganella acropodia		X							
Trigonopyxis arcula		X						X	
Trinema enchelys	X	X	X			X			
Trinema lineare	X								

*A mixotrophic species, which can also photosynthesise using endosymbiotic algae (see Figure 52). Modified from Wilkinson & Mitchell (2010), with additional data from Maillard *et al.* (2025). In constructing the original version of this table we made considerable use of information from Gilbert *et al.* (2000), which is still a useful paper for anyone who reads French and is interested in the food of testate amoebae.

from dead organisms back into the food chain.[20] For predators the general rule in food chains and food webs is that larger-bodied species tend to occur higher in the chain, as organisms usually prey on things smaller than themselves.[21] The idea that testates mainly feed on bacteria fits with this generalisation, but some can feed on larger organisms such as nematode 'worms'. For example, both *Apodera vas* and a *Difflugia* species have been recorded feeding on a range of nematode species in samples of leaf litter from native New Zealand forests.[22] These relatively large testate species consumed the 'worms' rather like someone sucking up spaghetti.

For a smaller testate, such as *Cryptodifflugia operculata* (at most 20 μm long) this 'spaghetti sucking' approach is not an option. In the more familiar world of large mammals in TV documentaries, we are used to seeing predators that hunt in packs or prides overpowering larger prey by working together. *C. operculata* seems to do something similar with nematodes.[23] Typically, one testate grabs the nematode by its tail and sticks it down to the sediment. Others now pile in to attack the partly immobilised 'worm', often consuming their prey within half a day or so. The obvious question is: has this amoeba evolved pack hunting as a discrete strategy for attacking prey too large for an individual to handle, or are the other amoebae making chance use of the immobilised nematode? The same question used to be raised about group hunting in Lions *Panthera leo* – were they really cooperating to hunt large prey, or did one lioness just accidentally scare the prey into the path of another member of the pride? Before he started his classic study of the behaviour of Lions in the Serengeti in the 1960s, George Schaller was inclined to think that accident was a more likely explanation than deliberate group hunting. However, he changed his mind after studying Lions in the wild.[24] The fact that *C. operculata* regularly uses the 'grab the worm by the tail first' approach, along with the suspicion that it then summons other testates via chemical signals, suggests that we are looking at pack hunting by amoebae – analogous to what we see in Lions or Wolves *Canis lupus* when they are hunting prey larger than themselves.

That amoebae can appear to show pack hunting behaviour is clearly fascinating, but there is a wider question of how important this is in the context of soil ecosystem processes. Nematodes are an important part of soil ecosystems, often being very numerous, with different species feeding on a wide range of other organisms – and, because of this, their diversity is often used as an indicator of soil 'health'.[25] Stefan Geisen and colleagues also showed, using molecular methods, that *C. operculata* was very common in soils, representing up to 4 per cent of the active protist community in the soils studied.[26] This suggests that it is common enough to play an important role as a predator of nematodes in soil food webs. So its pack hunting may influence soil ecosystem processes such as nutrient cycling through its effects on nematode populations.

This raises the obvious question of how important testates are as a group in the cycling of nutrients in soils.[27] The answer depends on multiple factors, such as the density of testates in the soil, and the rate at which material from testates is made available to other organisms. One way this can happen is by predation of the testates by other organisms – such as other protozoa and small soil animals. However, there is almost no quantitative information on rates of predation on testates, making it impossible to know how important this is. Indeed, it would be very hard to collect such data in a systematic way. Another potentially important role is to create hot spots of nutrients in the soil, as when testates die the nutrients in their cell contents and test are released back into the soil. These hot spots of nutrients could be an important resource for plants and other organisms. It's actually rather rare to see broken tests in soil samples, suggesting that when most tests start to break down they do so very rapidly (Figure 96).

Unfortunately, there are relatively few studies that try to quantify test breakdown, and so the rate at which nutrients are released from the test back into the soil is poorly unknown. Indeed, still the most detailed study available, by Lousier and Parkinson, was published way back at the start of the 1980s.[28] This study involved a series of experiments on the disappearance of empty tests

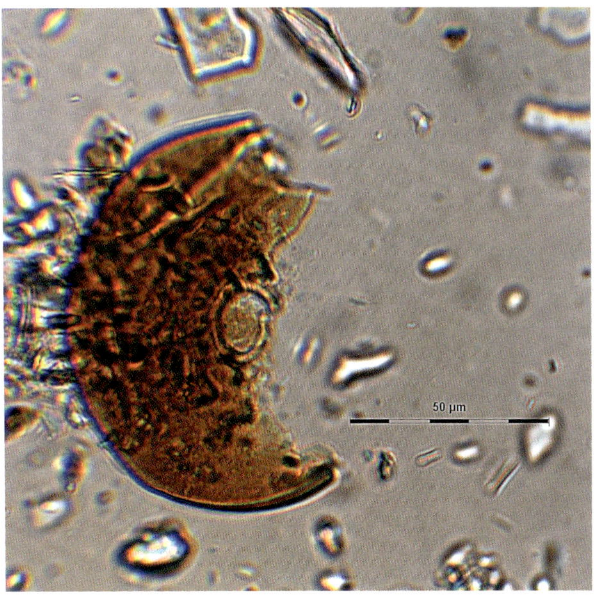

FIG 96. A broken test of an arcellinid testate amoeba (*Galeripora* sp.) from a moss sample from the Swiss Alps. It is unusual to see partly broken-down tests of most testate taxa, presumably because they fall apart very quickly in soil once they start to break up. However, broken *Galeripora* are reasonably common, as they seem to break down more slowly than many other testates.

in various soils, finding that for those made out of silica-rich plates 74–97 per cent of the tests had gone after one week. For species with tests made out of mineral particles the figure was 42–68 per cent lost after one week. Temperature seemed to have little effect on rate of breakdown, but soil moisture content was important – with tests surviving for longer in drier soils.

In the light of more recent studies, Lousier and Parkinson's finding that testates with tests built of secreted silica-rich plates break down quickly is particularly interesting (Figure 97). Although the silica cycle is not given much space in most general biology and ecology textbooks, which tend to concentrate on the carbon and nitrogen cycles, it's very important on a global scale. On a geological time scale the weathering of silicate rocks is a major mechanism for removing carbon dioxide from the atmosphere, while on shorter time scales the movement of silica from soils to water is important for the growth of diatoms. These form an important link in the carbon cycle, as dead diatoms sink to become preserved in marine and lake sediments, so taking nutrients, including carbon, from the water and entombing them in sediments.[29] Studies on the potential role of testate amoebae in the silica cycle (mainly the euglyphid testate amoebae) are very recent, only really starting in the first decade of the twenty-first century.[30] This work has shown that the amount of silica moving through the tests of silica-rich testate amoebae is similar to, and may possibly exceed, the annual uptake of silica by trees. When they die, these amoebae are creating hot spots of available silica in the soil for plants, and other organisms, to use. As silica is important in the growth and stress resistance of many plants this is likely of general importance.

FIG 97. The test of *Quadrulla symmetrica*, with its characteristic square silica-rich plates (technically referred to as idiosomes). Silica is also known as silicon dioxide (SiO_2). This species is typically 70–85 μm long, and is the most common species in this genus – typically found in wet mosses (Kosakyan *et al.*, 2025). *Quadrulla* is a relatively rare example of an amoeba in the Arcellinida with a silica-rich test; most such testates are in the Euglyphida (e.g. Figure 57). (Ferry Siemensma)

There are also good reasons to guess that the amount of silica associated with amoeba tests is currently being underestimated, so they may be even more important than we currently assume. While it's possible to estimate the amount of silica in intact tests in a particular soil,[31] much may be present in the form of tiny plates from decomposed tests – and it's much harder to estimate the numbers of these. Many people who have tried studying this area suspect that they may be very common. If so, the amount of testate-derived silica in many soils may be very high. As a field of research this is only around 20 years old, and there is much that we currently don't know.

PHOTOSYNTHESIS IN THE SOIL?

The general assumption has been that photosynthesis by algae in soils is of no importance, except in a few rather special habitats. The reason seems obvious – algae need light for photosynthesis, and except on its surface the soil is dark, and in many habitats even the soil surface is heavily shaded by plants. In addition, algae growing on soil are mainly limited to a two-dimensional surface, while larger plants may have many layers of leaves for photosynthesis. However, as with so many things that seem obvious, better data have shown that things are rather more complicated than previously thought.

Under some conditions the importance of soil algae has been long appreciated. The habitat where there is a long history of taking photosynthesis by soil algae seriously is the arid-land biocrusts. These were briefly described in Chapter 2, and are often a mix of mosses, lichen, cyanobacteria and eukaryotic algae (see Figures 22 and 23). Because of the arid conditions, larger plants are often few and far between, so lots of light reaches the soil surface. Another habitat where lots of light can get to the surface vegetation is on many peat bogs, which when in good condition often have a covering of bog-moss (*Sphagnum*) species. Over a century ago, in one of the founding books on plant ecology, Eugenius Warming noted cyanobacteria living in quantity in bog-moss viewed with a microscope, and wondered if they were important. He was unsure and concluded that 'The present state of our knowledge does not permit us to define the exact nature of the symbiosis in every case.'[32] It's now clear that such algae are potentially very important in the functioning of peatlands, as will be discussed in more detail in the context of climate change in the next chapter.

What about the role of algae in soils in general? Much of the recent research on this topic has been led by Vincent Jassey. As part of these studies, Vincent and seven colleagues attempted to estimate the contribution of soil algae to the global carbon cycle, using data from 203 locations around the world compiled from previously published studies – with the largest amount of data coming from European studies.[33] They estimated that on average there are around 5.5 million algal cells in each gram of dry surface soil (with the highest values found in some croplands, while broadleaved and mixed forests had much lower numbers). An obvious question is: how does this compare with larger plants when it comes to the assimilation of carbon from the atmosphere via photosynthesis? Jassey and colleagues estimated that these algae were taking up around 6 per cent of the amount of carbon assimilated by the growth of more familiar terrestrial plants – a large enough effect to be worth including in your calculations when considering land-based efforts to mitigate carbon emissions involved in global heating. It's long been known that marine algae can take up important amounts of carbon, but until recently most people assumed that terrestrial algae had such a small effect that they could reasonably be ignored in calculations about carbon take-up.

This global big picture is crucial for understanding the importance of soil algae in ecosystems, but it's often easier to put this in context at a much more local scale. The example most familiar to me is provided by the soil diatoms at Mere Sands Wood nature reserve north of Liverpool, as some years ago I studied them along with my then Masters student Angela Creevy, and my colleague Jane Fisher (who was the diatom expert). Despite it being a relatively limited study (all the soil samples were collected over just two days) we identified 52 species of diatoms in the soil samples – for comparison, we also found 42 species of testate amoebae in these soils. These soil diatoms are playing a similar role to the testate amoebae described above: the diatom's silica-rich frustule creates a hot spot of silica in the soil as it breaks down, so having a role in the silica cycle.

One of the things that made the reserve at Mere Sands Wood so useful for such a study is that it is a mix of habitats. We found diatoms not only in grassland soils but also under birch woodland, and perhaps unsurprisingly we found the most species in the wetter, mossy habitats. They were, however, absent from soils under dense pine plantations, and also under Rhododendron *Rhododendron ponticum* agg. (Figure 98), where presumably lack of light made conditions unsuitable for photosynthesis. The global survey of soil algae described above also found more algae in grassland soils than in deciduous woodland, and found much lower rates of algal photosynthesis under coniferous forest than broadleaved forest (be that deciduous or evergreen). It seems that diatoms, and other soil algae, can do well under all but the darkest woodland canopies.

The Mere Sands Wood study also highlighted another well-known aspect of soil diatoms. Many of these diatoms were quite small by the standard of aquatic diatoms (Figure 99). There are good reasons for this: in the 1940s, in an

FIG 98. Rhododendrons *Rhododendron ponticum* agg. at Mere Sands Wood, a Lancashire Wildlife Trust nature reserve. There were plenty of soil diatoms under areas of deciduous woodland, but none from the very deep shade under Rhododendron bushes. For more details of the diatom (and testate amoeba) studies at Mere Sands Wood see Creevy *et al.* (2016) and/or Wilkinson *et al.* (2017).

important early research paper on British soil algae, John Lund speculated that many soil diatoms were small because this allowed them to function in thin water films, so surviving better in soils than would the larger cells of the diatoms more often seen in rivers and lakes.[34] These ideas still seem very plausible today, as this small size would make it easier for diatoms to absorb water and nutrient from the water film (this follows from the discussion of surface-area-to-volume ratios in Chapter 3).

As well as contributing to carbon storage in soils, and in the case of diatoms being involved in the silica cycle, soil algae have several other ecosystem roles. Many cyanobacteria can fix nitrogen, that is take nitrogen from the air and convert it into compounds more readily usable by most organisms. When Warming was looking at cyanobacteria in bog-mosses at the end of the nineteenth century, nitrogen fixation was one of the processes he was wondering about. Perhaps the most basic ecosystem process, and certainly the one we learn about at a young age in school, is providing energy for organisms higher up in

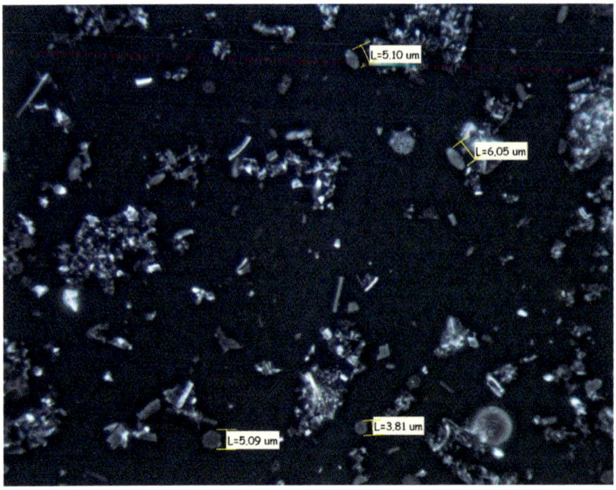

FIG 99. Small soil diatoms from Mere Sands Wood. In this case they are mainly between 4 and 6 μm long. (Angela Creevy)

the food chain. For example, a study in Ireland, using carbon dioxide 'labelled' with a distinctive stable isotope of carbon so it could be followed along the food chain, found that several soil invertebrates were gaining useful amounts of their energy from soil algae. In simple experiments using soils from both coniferous woodland and an agricultural field, the springtail *Ceratophysella denticulata* was found to be obtaining up to 17 per cent of the carbon in its body from soil algae (the value for the earthworm *Allolobophor chlorotica* was 3.0 per cent). This suggests that these algae provide a small, but important, food source for soil invertebrates.[35] However, when thinking of algae it is aquatic habitats that normally come to mind, and it's these freshwater ecosystems that I consider next.

A VIEW FROM WINDERMERE: FRESHWATER PROTISTS

John Lund, whose soil work on algae was described above, spent much of his career based at the Freshwater Biological Associations (FBA) laboratories on the shores of Windermere (Figure 100). For much of the twentieth century the FBA labs were a major centre for research on aquatic freshwater protists in Britain. In 2021, based on the many decades of work, Pentecost and Haworth were able to list 2,196 taxa (both species and varieties) of freshwater algae from Cumbria – spread across 13 phyla.[36] Windermere has been particularly well studied and provides a good example of the way in which different algae can vary in a lake over the course of a year. Stephen Maberly and colleagues have used a long run of FBA data, from 1945 to 1977, to provide a summary of how the algal community

FIG 100. View from Fell Foot looking north along Windermere, in the English Lake District, showing the southern part of the lake. As home to the Freshwater Biological Association this lake, and others nearby, became key sites in freshwater research in Britain, much of it focusing on algae and other protists. Since the start in the 1930s its labs have been based successively at three different locations on the west shore of the lake (and have recently moved to new offices southeast of Windermere). It continued as a site of world-leading research until funding changes in the late twentieth century. Long-term studies of Windermere, and other nearby lakes, was continued by the Centre for Ecology and Hydrology, but no longer based at the FBA labs. As well as scientists directly employed by the FBA, Windermere was also home to groups with other sources of funding (e.g. the Institute of Freshwater Ecology). Work by several scientists based at the FBA Windermere labs has already been described in previous chapters – such as the studies at Priest Pot led by Bland Finlay, and Kathleen Atkinson's work on ducks as dispersal agents for freshwater algae. Maberly *et al.* (2018) provide a useful overview of the long-term freshwater research in the Lake District.

changes in the South Basin of Windermere over the course of a year (Figure 101).[37] These data are particularly valuable, as all the algae were identified either by John Lund himself, or by junior colleagues of Lund's working under his supervision. This reduces problems with different people assigning algae to different species – a potentially serious issue with large data sets of microscope-based protist identifications involving many different scientists potentially using different criteria to identify the species.

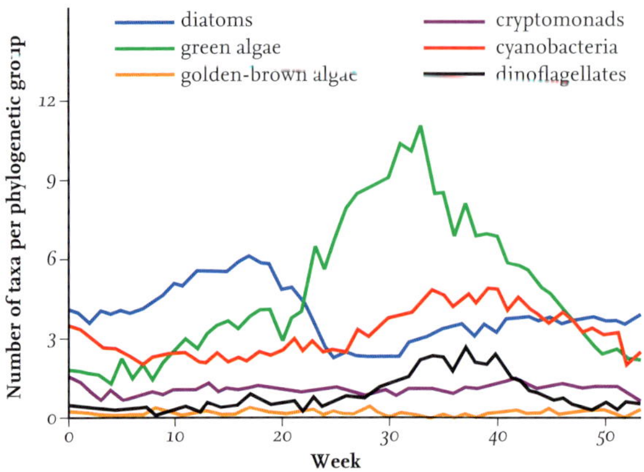

FIG 101. Average pattern of species richness in different phyla of algae in the South Basin of Windermere plotted using data collected between 1945 and 1977 (during this period algal samples were usually collected on a weekly basis). Week '0' is the start of January. Scientific names of these phyla are given in Table 2 (page 76). (Graph redrawn from Maberly *et al.*, 2022 – originally published under a Creative Commons licence, CC BY 4.0)

These data show that diatoms were typically the most species-rich group early in the year, before declining over summer, and then showing an increase in species again later in the year. As the diatoms declined the number of species of green algae rose, reaching a peak in the summer. The next most species-rich group were the cyanobacteria, again showing a peak over the warmer summer months and declining during the autumn (on occasion cyanobacteria can become extremely common in warm conditions, a topic discussed in more detail later in this chapter).

As well as analysing this long run of data from Windermere, Maberly and colleagues looked at data from 20 Lake District waterbodies sampled every five years between 1995 and 2015, an ongoing project known as the 'Lakes Tour'.[38] In each tour year samples are taken in winter (January), spring (April), summer (July) and autumn (October). This wider sample showed similar patterns to the earlier weekly Windermere data set. The researchers concluded that the higher algal species richness seen over the summer was linked to higher water temperatures, but also found evidence for increasing competition between species in summer. Typically protozoans, such as ciliates, also increase in numbers over the summer as more food becomes available.

Given that warmer summer water attracts not only human swimmers but also algae, what is going on with diatoms? Why does diatom species richness peak early in the year, then decline before a partial recovery after the summer? It's not just species richness, but numbers of individual diatoms too. Classic work by John Lund on Windermere diatoms, especially *Asterionella formosa*, showed that numbers of individuals in a given species also peaked early in the year, and his suggestion that silica supply was a large part of the explanation still holds.[39] As described in more detail in Chapter 4, diatoms have silica-rich frustules (Figure 102). This means that as diatom numbers build during the spring they are using up the silica supply in the water, and as silica declines so do diatom populations. So although the green algae, and other groups, are able to expand in the warm summer water the diatoms struggle as the key chemical needed to make new frustules is in short supply. One of the reasons that diatom numbers can increase again in autumn is that some of the silica in the water is replaced in runoff from surrounding soils – where, as described above, soil diatoms and testate amoebae have helped make this element available. Clearly there will be many other processes involved, such as temperature and competition with other algae; however, changes in silica levels are one of the key drivers explaining changes in diatom abundance and species richness over the course of a year.

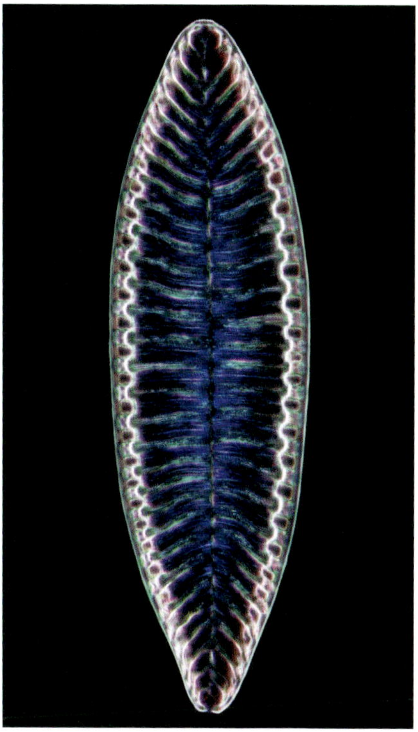

FIG 102. The strikingly beautiful silica-rich frustules of the pennate diatom *Iconella biseriata* (formerly in the genus *Surirella*). This diatom has been found in many of the larger Lake District lakes – including Windermere – along with 10 of the smaller lakes and tarns (Pentecost & Haworth, 2021). (Alex Hyde)

How typical are these patterns? Malham Tarn in Yorkshire is another well-studied lake, but very different in its chemistry from lakes such as Windermere. The tarn is on limestone and is an unusually large marl lake, with a high level of dissolved calcium in the water. There are only four surviving Cumbrian examples of marl lakes, all associated with limestone, so the water chemistry at Malham is very different from the main Lake District lakes and tarns on the harder rocks of the central fells. Marl lakes are of conservation interest in part because some species of stonewort (see Figures 38 and 39) are characteristic plants of many of these

FIG 103. The Malham Tarn weather station, with a mist-shrouded tarn behind. Continuous weather records from here start in 1948 – although there were some earlier records from the late nineteenth and early twentieth centuries (Lee, 2015). See Figure 104 for a comparison of some of the mid-twentieth-century data from this station with diatom abundance.

lakes, including Malham Tarn. Along with the macroscopic stoneworts, the tarn has many microscopic algae, and altogether over 270 species of algae have been recorded.[40] These algae have been well studied over the years, in the past often involving scientists associated with the FBA. As there has also been a weather station on its shore since the mid-twentieth century, there is good climatic data available to help interpret seasonal patterns in the algae (Figure 103). John Lund also studied the Malham diatoms, and found broadly similar patterns to those he had found in Windermere – with peaks early and late in the year, and low numbers during the summer when silica levels were low (Figure 104).

Silica levels are important to diatoms because silica is needed to form their characteristic frustules. Other nutrients are more widely important for freshwater algae. The two key elements are phosphorus and nitrogen; of these, phosphorus is often the most important in limiting alga numbers. If phosphorus is plentiful, but nitrogen is in short supply, this tends to favour cyanobacteria – as many species are able to fix nitrogen. Famous experiments in Canada in the 1970s led by David Schindler showed that experimentally reducing the amount of nitrogen while maintaining phosphorus levels caused cyanobacteria to dominate.[41] In most freshwater and soil habitats this nitrogen fixation takes place in specialist cells called heterocysts (see Figure 65 – the heterocysts are the larger circular cells that can be seen in some of the algal filaments), although in the ocean most cyanobacterial nitrogen fixation takes place without the need for these cells.[42]

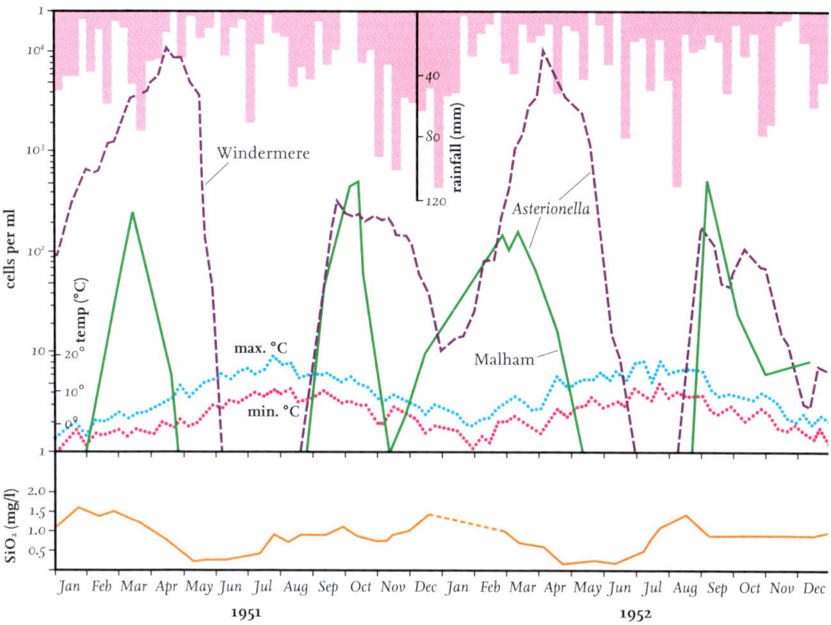

FIG 104. Density of cells of the diatom *Asterionella formosa* in Malham Tarn in the early 1950s. The Malham data are compared with similar studies in Windermere. Note also the bottom curve in the graph, which shows the concentration of silica (SiO_2) in the Malham Tarn's water. (Graph modified from Lee, 2015, based on studies by Lund, 1961)

Changes we have made to the environment, such as pollution from agriculture and sewage disposal, have increased the number of waterbodies that have plentiful supplies of phosphorus, leading to increases in cyanobacterial algal blooms (Figure 105). Warmer weather also increases the likelihood of such blooms, so a combination of nutrient runoff from human activities and a warming climate is tending to make such blooms commoner and more of a problem. Water temperature and nutrient levels interact in producing blooms, and it's been calculated that a nutrient reduction of around one-third would be needed to offset the effects of a warming of 1 °C.[43] Algal blooms are a particular worry, as some of the species involved release toxins into the water. Such is the interest, and importance, of these blooms that there is an entire scientific journal called *Harmful Algae* devoted to research in this area.

My description of freshwater algae in this section has been based on studies using microscopy, which need the organisms to be common enough to be seen in a water sample. In the last few decades, however, it has become apparent

FIG 105. Algal bloom in a Trafalgar Square fountain during a London heat wave – the water colour suggests that cyanobacteria are mainly responsible. The nutrients likely come from urban air pollution, and perhaps also the square's famous pigeons. Cyanobacteria can produce a range of chemicals especially toxic to mammals (including ourselves, our pets and livestock), making such blooms a significant health risk (Dow & Swoboda, 2000).

that many microbial species are so rare that they are effectively invisible to microscope-based approaches. These microbes form the so-called rare biosphere.

THE RARE BIOSPHERE

Lakes and their microbes nicely illustrate an important ecological problem. Consider the case of a simplified 'mathematical' lake. This small lake is square in shape and has sides 100 m long, and it is of uniform depth, just 1 m (its size and shape are obviously chosen to keep the maths simple). Imagine a species of microscopic alga that is relatively uniformly distributed through the lake's water, and has a total population of 1 million cells. A million individuals sounds like a large population, but on sampling such a lake there would be around 100 cells in every cubic metre of water sampled – so rare that they would be very hard to find by microscopy. If the algae were even moderately clumped in their distribution, many samples would contain none at all. This imaginary lake was described by the ecologist Evelyn Hutchinson in the mid-twentieth century, and he concluded that such an algal species 'would be far too rare ever to find by ordinary techniques'.[44] 'Ordinary techniques' at the time consisted of concentrating the algae in a water sample by some means (see below) and then viewing them with a microscope.

Rare species are potentially very important in ecology, as most species are rare – so most of biodiversity consists of rare species. Typically, if you survey a particular group of organisms at a particular location there will be a small number of common species that dominate the assemblage and a long tail of rare species – Figure 106 shows a beetle example. It's hard enough surveying rare larger organisms, be they beetles or mosses (remember – these are unusually large organisms, even though most naturalists would describe them as small), and trying to quantify rare microbes has been very difficult. In Hutchinson's words, just 'too rare ever to find ...'

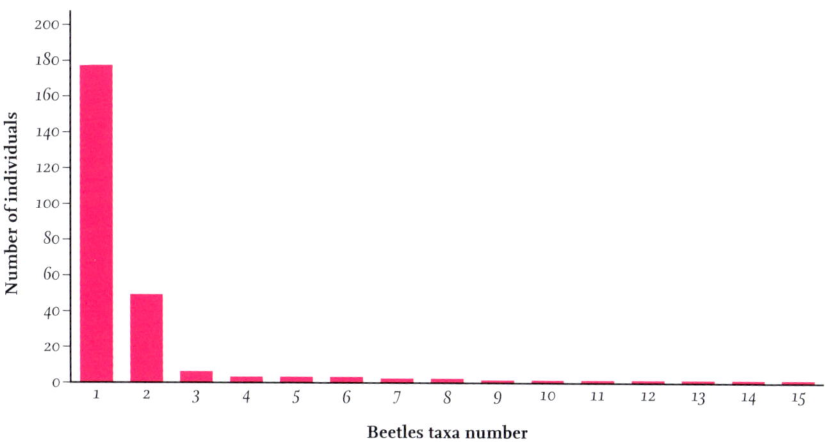

FIG 106. Most species are relatively rare, or as Gaston (2023, p. 20) puts it, 'among species the state of commonness is itself rare'. The graph illustrates a non-microbial example of a type more familiar to most naturalists. It shows the number of individuals of 15 taxa of beetles caught in a pitfall trapping study around the margins of a small pond on Llanbedr Hill in mid-Wales (data from Sanderson & Wilkinson, 1989). Most of the individuals caught came from just two taxa – the most common species were *Nebria salina*, and the staphylinid genus *Quedius* (species-level identification in this difficult genus was beyond us).

The 'ordinary techniques' that Hutchinson had in mind were various ways of filtering algae out of the water, or other methods of concentrating the algal cells before putting the sample on a microscope slide to identify and count the cells. A longstanding method is to use a plankton net, with a filter fine enough to trap the microbes of interest. These methods are still in use (Figure 107), but recent technological changes have made it much more practical to go looking for Hutchinson's very rare microbes. As will be clear from many examples already discussed in this book, the largest change has been the introduction from molecular biology of methods that allow people to look for the genes associated with a particular organism, rather than having to find the microbe itself using a microscope. Away from microbial ecology, many naturalists are aware that environmental DNA is now commonly used for surveying ponds for amphibians, along with a growing number of other applications in ecological surveys.

Not only has the sophistication of the technology improved greatly over the last few decades, but its cost has reduced spectacularly too. This is a key point in making these methods available for widespread use by ecologists, who are usually not that well funded. For example, in 2004 Ford Doolittle calculated that the cost of sequencing a million base pairs of DNA had dropped by 100,000-fold over just 15 years.[45] In part, ecology is benefiting from the spin-offs from medical research – including the Human Genome Project – which started in 1990 and has driven major advances in the technology, making it much cheaper to use. This has enabled ecologists to employ metagenomics approaches which can sequence (i.e. identify) all the DNA or RNA in a sample, allowing the identification of the microbes through their genetic material in a particular sample.

OPPOSITE: **FIG 107.** Sampling for algae and other microbes in freshwater habitats in southern China. (a) Sampling for plankton in Bantou Reservoir. To the left, a plankton net is just being pulled up after being used to sample microbes in the water column. Behind can be seen the outflow pipe attached to a pump that can be used to sample water from different known depths, and in the foreground is a probe for recording aspects of the environment – such as temperature, pH and dissolved oxygen content of the water. This reservoir formed part of the cross-China study of the bacterial rare biosphere described in the main text. (b) A bridge over the Houxi River in Xiamen City is being used as a stand-in for a boat – allowing us to easily sample water from the middle of the river. The sampling container (centre of the photograph, held by Alain Isabwe) is designed to be lowered into the water to a particular depth before being opened to sample the water – a more common approach than using the pump and tubing shown above. The samples are then taken back to the laboratory to extract the microbes for study by either microscopy or molecular methods.

a

b

For the origins of this metagenomic (or microbiomic) approach to looking for Hutchinson's 'too rare ever to find' microbes, we need another brief excursion into marine microbiology. The first use of the term 'rare biosphere' seems to have been in a 2006 research paper by Mitchell Sogin and colleagues.[46] They pointed out that given the high numbers of bacteria found in seawater it seemed odd that relatively few species were known, and used molecular methods to show that while a small number of types dominated a marine water sample (these were the ones you would find using a microscope), thousands of much rarer types could be identified in these samples using molecular techniques. They named this long list of bacteria invisible using microscopy 'the rare biosphere', and this work led to multiple similar studies, often in marine systems and concentrating on bacteria.

In the 20 years since the publication of this first 'rare biosphere' paper the technology has been steadily improving, making such studies easier to carry out. Ten years after the first marine paper I was involved in one of the first large-scale studies of the bacterial rare biosphere in freshwater – looking at 42 lakes and reservoirs scattered across China.[47] We found that fewer than 2 per cent of the bacterial types were abundant, while 72 per cent were rare, so rare types make up most of the bacterial diversity of these lakes. One of the more interesting results of this cross-China study was that we found ecological patterns in the rare bacteria – the rare biosphere was not just a random assemblage of small numbers of microscopically invisible cells. Some of the patterns in the rare biosphere appeared to be correlated with various local and regional environmental factors. This is relevant to the idea that 'everything is everywhere' discussed in Chapter 5. If this were the case, one would expect the rare biosphere to include everything and so not be correlated to local environmental conditions (always assuming that there isn't an even rarer biosphere, still invisible to our current molecular methods).

Although the majority of these rare biosphere studies have so far concentrated on prokaryotes, there is a growing number of studies of other microbial groups, and also from a wider range of habitats including soil.[48] For example, there are now multiple studies of the rare biosphere for marine and freshwater ciliates. These show a large diversity of rare types based on their genes, although it's still unclear if these data really represent that many actual species, and if so how many there are, and what their ecological functions are.[49]

The microbial rare biosphere is interesting for several reasons. First, it shows that the same pattern we see in larger organisms, of a small number of common species and a much larger number of rare species, is also found in the microworld. In both macro- and microworlds these rare species may contain future common species if environmental conditions change to favour them (think of them as analogous to a seed bank, containing species that may be important in the future). In addition, in the case of the microbial rare biosphere, especially among the prokaryotes (including cyanobacteria), there is another potential importance. Because of the horizontal transmission of genes between microbes, these rare microorganisms can be seen as a gene bank, containing genes that could be transferred to commoner species. So this rare biosphere may function as an important reservoir of microbial diversity – as both a 'seed bank' and a gene bank.

MICROBES ARE CRUCIAL TO ECOSYSTEM PROCESSES

Microbes form a substantial part of the genetic and biochemical diversity of life on Earth, and also contribute large amounts of biomass – because although by definition they are very small, and each one weighs very little, they are also present in huge numbers. It's much harder to estimate what proportion of species are microbial, in part because we are very unsure how to define species in microorganisms; this is especially true for the prokaryotes, where the transfer of genes between individuals can be common. Given their diversity, and their large numbers, it's unsurprising that they play crucial roles in ecosystem processes. In the past most ecologists have overlooked the importance of microbes, and tended to focus on larger organisms. But this is starting to change, with molecular methods making it easier to collect large relevant data sets for studying microbial ecology – although we still have much to learn about how to analyse and make sense of this rapidly growing mass of data. This will likely form a very important research area over the next few decades.

The rapid rate of change in our understanding is illustrated by several examples in this chapter. Ideas of the microbial loop only started to coalesce in the early 1980s. Our understanding of the importance of soil algae, or testate amoebae and diatoms in the soil silica cycle, only took off in the last couple of decades (Figure 108). A key change has been in the increasing availability of powerful techniques from molecular biology. Among many other contributions to microbial ecology, these methods have made at least part of the rare biosphere 'visible' for study. Although Evelyn Hutchinson speculated about an aquatic rare biosphere in the mid-twentieth century, it's only in the past 20 years or so that we have had the technology to investigate it in any detail.

The key role of the microworld in the functioning of ecosystems means that microbes are potentially very important when it comes to understanding

FIG 108. A collection of marine diatoms. The silica cycle in soils is important in releasing silica to lakes, rivers and ultimately the sea – where it is crucial for diatom frustule formation. Today, diatoms play an important role in both marine and freshwater ecology, being eaten by many other organisms. However, this importance is geologically quite recent, as diatom fossils are not found until the Early Cretaceous – around 120 million years ago (Bryłka *et al.*, 2024). They appear even later in freshwater ecosystems, first appearing in the Late Cretaceous, but only becoming common later. The rise of diatoms provided new food sources for small aquatic animals such as waterfleas (Cladocera), making the diatom-free ecology of lakes at the time of the dinosaurs rather different from those we see today (van Damme & Kotov, 2016). (Alex Hyde)

the planetary ecosystem, and human effects on it. This planetary ecosystem is sometimes referred to as Gaia, and one of the things that marked out Gaia theory as different from more traditional ecological approaches to the planet was the importance it gave to the role of microbes. In part this was because of the involvement of Lynn Margulis in the early development of these ideas.[50] The role of microbial processes in understanding the many human effects on the functioning of Gaia forms the subject of the final chapter.

CHAPTER 9

Conservation and Global Change

The path zigzags up the steep hill, gaining height by stealth, with ferns prominent on its shadier margins. Towards the top I stop to look back over the village through a gap in the trees. On my phone I have a copy of a photograph of this view taken around 1900, and I have a small bag of camera equipment with me to retake the scene. My bag contains a smallish SLR camera and a couple of lenses, so weighs only around 2.5 kg – but some 125 years ago the early photographer (one or both of the Kearton brothers) would have carried much more equipment up to this spot. As well as their large camera, and the substantial wooden tripod needed to support it, there were also the heavy glass plates that were the photographic 'film' of the time. I can vouch for the weight of those wooden tripods, as in the early 1980s when I was learning surveying techniques a few were still in use.

The village below is Selborne in Hampshire, and the hill I am standing on is known as 'the Hanger'. In the eighteenth century the village and its surrounds were the setting for Gilbert White's *Natural History of Selborne* – arguably *the* most widely read book on British natural history. It was first published in the late eighteenth century, and is still in print today.[1] The pioneering wildlife photographers Richard and Cherry Kearton produced an edition in 1902 illustrated by their photographs, including both views taken around Selborne and images of some of the species White wrote about. Their view of 'Selborne from the Hanger' looks down on the village around the church, and out across the fields beyond. Comparison with the view today shows that trees have increased, obscuring some of the buildings in the village – looking past the church, the hedge lines of 1900 were treeless, while today they have acquired several trees (Figure 109). Overall, the impression is of surprisingly little landscape change,

FIG 109. (a) View over Selborne from the Hanger, from the edition of *The Natural History of Selborne* illustrated by Cherry and Richard Kearton, first published in 1902 by Cassell. This plate comes from a later edition published by Arrowsmith in 1924, and shows the wider landscape more clearly than in the version of this view used in 1902 (however, comparison of the two shows they must have been taken at the same time – not 20+ years apart – so this photograph dates from around 1900). (b, opposite) A similar view taken in 2025; the church is now obscured by trees, but its graveyard can still be seen.

except for a general increase in larger trees. Comparison of the 1900 photograph with maps and illustrations from the eighteenth century also shows an increase in trees; for example, the zigzag path on the Hanger, which was created by White,

b

ran to the side of the woodland rather than through the trees as it did in 1900 and still does today.

These apparently limited changes around Selborne over the last few hundred years are deceptive. Humans are changing the climate and ecology of the planet on a scale that is hard to grasp while looking out over an English village which, at least when viewed at a distance, has looked superficially similar for some centuries. That is, of course, part of the problem. As you look around, the extent of the changes we are making is often hard to see, and so difficult to take as seriously as we should.

To give a sense of the scale and urgency of these changes, when giving talks on ecosystem ecology, I sometimes perform the opening paragraph of Bruno Latour's book *Facing Gaia*. Latour was one of the more significant academics in the humanities in the later part of the twentieth and start of the twenty-first century. Much of his later writing focused on environmental issues, which he considered to be '*a profound mutation in our relation to the world*' (the italics are his own).[2] Bruno's lecturing style could be somewhat theatrical, and *Facing Gaia* started as a series of lectures given in Edinburgh – hence my description of 'performing' the paragraph. It's perhaps best read aloud – and certainly overacting a bit:

> *It doesn't stop; every morning it begins all over again. One day, it's rising water levels; the next, it's soil erosion; by the evening, it's the glaciers melting faster and faster; on the 8 pm news, between two reports on war crimes, we learn that thousands of species are about to disappear before they have even been properly identified. Every month, the measurements of carbon dioxide in the atmosphere are even worse than the unemployment statistics. Every year, we are told that it is the hottest since the first weather recording stations were set up; sea levels keep on rising; the coastline is increasingly threatened by spring storms; as for the ocean, every new study finds it more acidic than before.*[3]

To start to put numbers to this, when White was writing his book in the second half of the eighteenth century, atmospheric carbon dioxide was around 280 parts per million (ppm). When the Keartons took their photograph of the view from the Hanger it was 296 ppm, but when I took my version of that photo at the start of June 2025 it was 429 ppm. From the discussions in this book so far it should be clear that the microworlds are crucial to the working of ecosystems, and so will be intimately bound up with the playing out of all the changes we are making to global ecology, as exemplified by the increasing concentration of greenhouse gasses in the atmosphere.

This is a huge subject, which could easily make a very long book in its own right. In this chapter I introduce the topic via a number of stories – with testate amoebae and diatoms taking the leading roles, along with a supporting cast of green algae, slime moulds, cyanobacteria and others. These problems are on a planetary scale, and so this chapter makes use of examples with a geographical spread wider than usually found in a New Naturalist volume, covering much of the rest of Europe and beyond.

To start, consider Latour's point that 'thousands of species are about to disappear before they have even been properly identified'. We are familiar with this crisis affecting the larger organisms that usually feature in natural history books. But what about microbes? We strongly suspect that most are as yet undescribed by science, but could they be becoming extinct?

DO WE NEED TO WORRY ABOUT THE CONSERVATION OF MICROORGANISMS?

The larger organisms familiar to most naturalists include many species that have been made rarer, or even extinct, by human actions (Figure 110). As described in the previous chapter, most organisms are rare – it's the really common species

FIG 110. In the non-microbial world, we are used to species becoming rare due to human actions, and even becoming locally or globally extinct. One example in Britain is the Eurasian Beaver *Castor fiber*, which is currently being reintroduced to many parts of the country, as here at the rewilding project at Wild Ken Hill in Norfolk. Their dam-building activities have attracted interest as potential flood control measures. The microbial ecology of beaver ponds is interesting, and important in the context of climate change. While the muds on the bottom of the pond can store carbon, the ponds also provide good conditions for methane-producing archaea living in their oxygen-poor sediments – and some of this methane escapes into the atmosphere, where it is a potent greenhouse gas. So alongside the many environmental benefits that are associated with beaver conservation and reintroduction programmes, there is also a potential downside from increased methane production (Whitfield *et al.*, 2015). Back in the 1980s Euan Nisbet (1989), building on work by T. E. Ford and R. J. Naiman, suggested that the then recovering Canadian Beaver *C. canadensis* population was likely important for methane production, with wetlands that have been altered by beaver activity producing substantially more methane than a beaver-free wetland. Euan was somewhat ahead of his time in taking an interest in beavers and methane; this is now a growing area of research, and there are currently multiple studies ongoing in an attempt to get a fuller understanding of the microbial ecology of methane and beaver ponds.

that are unusual. This means that many species naturally have relatively low population sizes at any one location, potentially making them vulnerable to environmental changes, and possibly local extinction.

Are we also causing extinctions in the microworld? The natural history of Britain has been studied in more detail than many other parts of the world – the fact that there are over 150 volumes in the New Naturalist series is one indication of the extent of this knowledge. Clive Hambler and colleagues used the large amount of historical natural history data from Britain to try and estimate extinction rates over the nineteenth and twentieth centuries.[4] Few microbial data were available for them to use, but they did manage to estimate British extinction rates for a small number of groups described in this book. These were lichens (which I have suggested form an accessible link between the macro- and microworlds), stoneworts (green algae, albeit large multicellular ones), and one group of protozoa, namely the plasmodial slime moulds (Table 4). From their results it does appear that Britain has been losing some larger 'microbial' species over the last couple of hundred years, possibly at a rate at least equal to that suffered by larger and more familiar organisms. However, it's almost impossible to say anything quantitative about extinction rates for the rest of the microworld in Britain – although, as has already been pointed out in this book, some of their habitats are becoming exceedingly rare, and with their reduction specialist microbes must also be close to being lost.

An obvious example of a microbial habitat under threat is provided by the persistent snow patches in the mountains, which are known to contain specialist algae and slime moulds. Back in the eighteenth century, during a cold period often called the Little Ice Age, these patches were much more extensive. For example, when Thomas Pennant – today most widely known as a correspondent

TABLE 4. Estimated extinction rates for Britain over the nineteenth and twentieth centuries (1800–1999). Data from Appendix C of Hambler *et al.* (2011). The only 'microbial' groups in this study were the slime moulds (which form visible structures during their life cycle), the stoneworts, which are green algae (albeit visible multicellular examples) and lichens, which although visible can be thought of as microbial symbioses. Data for more familiar, and well-studied, birds and larger plants are included for comparison.

Group	Extinction rate (as percentage of species in the group in 1800)	Hambler *et al.*'s confidence in their estimate
Plasmodial slime moulds	2.9	Low
Stoneworts	5.0	Very high
Lichens	0.8	Medium
Birds	2.4	Very high
Flowering plants and conifers	0.4	Very high

of Gilbert White but at the time a renowned writer on natural history and travel – toured Scotland in 1769 he reported perpetual snows on the Cairngorm summits, and ice on a high lochan even in late July.[5] The ending of this cold period during the nineteenth century, followed by recent human-driven rapid climatic warming, has brought this habitat to the edge of extinction in Britain. Another contender already described is the testate amoeba *Hyalosphenia papilio paynei*, which is only known from a very few bogs in Britain and Ireland (see Figure 61).

Outside Britain there are other examples. Consider the cyanobacterium *Nostoc* that Gilbert White used to see on the lawns and paths around eighteenth-century Selborne (see Chapter 7). The species he saw was likely *Nostoc commune*, a cyanobacterium with a very wide distribution. More recently it has been split into multiple species, including *N. flagelliforme* – which used to be considered just a variety of *N. commune*. *N. flagelliforme* has a rather distinctive cylindrical body and is found in arid habitats – such as in Spain. It is an example of a cyanobacterium of conservation importance; indeed, it is protected in China where it is considered endangered because of extensive harvesting as a culinary delicacy. In Europe it is apparently rare, and so potentially endangered too.[6]

Other algae may be in need of conservation as well. Diatoms are a good example, and in the 1990s Mann and Droop considered some diatoms to be very localised – such as species restricted to single lakes in both Africa and Russia.[7] They pointed out that the extent to which diatoms appeared endangered depended on your approach to defining species. The more you split diatoms into multiple species based on subtle differences in size and shape, the more endangered species you ended up with. In the three decades since that point was made, molecular methods have tended to lead to even more extensive splitting of microbes into multiple species (as in the case of the *Nostoc* cyanobacteria), increasing the number of species with very limited distributions. However, it is still the case, as Kent Redford recently pointed out, that 'many conservation organisations are largely uninterested in extending their work to microbes'.[8] In the few cases where 'microbes' have been treated as of conservation interest it's often groups such as stoneworts or lichens, organisms which cross the microbe/macrobe boundary.

Does this matter? It's possible that microbial conservation may be rather different from the conservation of larger organisms, and that we need to worry less about microbial extinctions than we do in the case of the more familiar multicellular animals and plants. If the longstanding idea that for microbial diversity 'everything is everywhere' were largely correct, then extinction of most microbes would seem very unlikely. As described in Chapter 5, despite many uncertainties, it's clear that not *all* microorganisms can be found globally, even in habitats that appear suitable for them (for another example, see the discussion of

Apodera vas, below). However, many microbes do seem to have wider distributions than we are used to seeing in larger organisms.

I first used the subheading for this section ('Do we need to worry about the conservation of microorganisms?') as the title of a book chapter I wrote almost 20 years ago.[9] At that time it was still unclear to me to what extent the 'everything is everywhere' idea might be correct. If true, at least to a first approximation, then extinction of free-living microbes might be quite rare.[10] However, even if 'everything is everywhere' were correct, extinction due to human actions might still happen if their global distribution was maintained by a balance between local extinction and recolonisation from elsewhere. If the changes we are making increase extinction, at the same time they will be reducing the number of locations from which vacant patches could be recolonised, and at some point

FIG 111. The Hottentot-fig *Carpobrotus edulis* (flowering in the foreground) is a South African species extensively naturalised in the Mediterranean and Atlantic regions of Europe. In Britain it is currently limited to the south, and is considered invasive in the Channel Islands and Isles of Scilly. It is 'perceived as a serious threat to native species', but currently limited by climate in Britain (Stroh *et al.*, 2023). The example shown was taken in Vamos in Crete, where the Mediterranean climate is similar to that in its native range.

extinctions start happening faster than recolonisations, and global extinction becomes possible.[11] The fact that we now have multiple examples of free-living protists with restricted distributions makes global extinction more likely than in a microworld where everything truly was everywhere.

Two key issues in the conservation of larger organisms are the effects of introduced species and climate change. For example, for plants an estimated 59 per cent of Britain's flora comprises introduced species – for comparison, the figures for two other European countries where comparable data are available are 33 per cent for the Czech Republic, and 21.5 per cent for Slovakia.[12] A small proportion of these introduced species compete with native plants to an extent that leads them to be considered problematic from a nature conservation perspective. In some cases, climate change is likely to make some introduced species spread in a way that may increase their potential for causing problems (Figure 111). Climate change can also alter the habitat of many species, as is the case for the declining snow-patch microbes. How these two themes of introduced species and climate change might affect the less familiar microworld is illustrated by a testate amoeba called *Apodera vas*.

THE STORY OF *APODERA VAS*

Apodera vas (Figure 112) is particularly interesting for several reasons. It is usually found in cool to moderately cold, damp or wet peatlands, or forests with mosses and high levels of organic matter in the soils. In summary, it likes a wide range of wet, and often mossy, terrestrial habitats. Britain abounds in such habitats, especially in the north and west, but the species has never been found in samples from Britain, or indeed anywhere in Europe. One of the reasons this testate has featured so prominently in the literature on protist diversity and

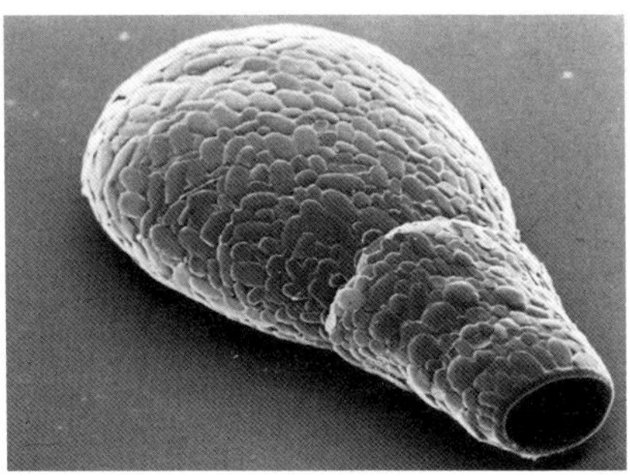

FIG 112. Scanning electron micrograph of a test of *Apodera vas* (called *Nebela vas* in the older literature). Note its distinctive structure, including the characteristic constriction where the neck of the test joins the body, although there is a strong suspicion that there may be several similar species making up this morphospecies (Bruni *et al.*, 2024). The test is typically around 150 μm long. This example came from New Zealand. (Wilhelm Foissner)

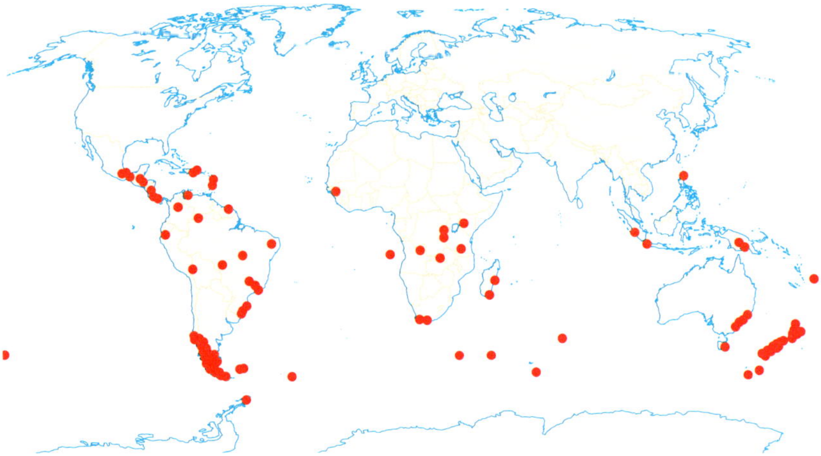

FIG 113. The current distribution of *Apodera vas*, based on 401 records of its occurrence around the world. (Redrawn from Bruni *et al.*, 2024; Creative Commons licence, CC BY 4.0)

biogeography is that it has a rather striking global distribution – being widely distributed in the southern hemisphere but only getting into a few low-latitude parts of the northern hemisphere (Figure 113). So *A. vas* provides a convincing example of everything *not* being everywhere, which is why its distribution has been of particular interest in debates about protist diversity over the past few decades.[13]

In many microbial examples it can be very difficult to be sure that a species is genuinely absent from an area – as they are by definition microscopic, and the number of relevant studies is often very limited. This is why *A. vas* has attracted so much interest, including being used as a very rare microbial case study in several editions of one of the leading textbooks on biogeography – introduced as 'the strange case of the testate amoeba'.[14] The key point is that this is a distinctive-looking testate (so very unlikely to be confused with other species), and most crucially it is absent from those parts of the world (Europe and eastern North America) which have been studied in the most detail. So we can have confidence that its absence from much of the northern hemisphere is real, rather than an artefact of inadequate data.

As discussed at some length in Chapter 5, there are two parts to the 'everything is everywhere' idea. First, that there is a global distribution (everything is everywhere), but also that a particular microbe will be found only where the conditions are right for it (the environment selects). In the original 2007 analysis of global *A. vas* distribution by Humphrey Smith and myself we just pointed out that its mossy, and sometimes forested, habitats were widely found across the northern hemisphere.[15] More recently, Bruni and colleagues were able to construct a statistical model to describe the climatic requirements of this testate – using data on the climate at its known locations to predict places with suitable conditions.[16] This suggested that many parts of the world where the amoebae were not found had a suitable climate – including Britain. These statistical relationships could also be used to predict the potential effects of future climate change on the availability of suitable habitats (Figure 114). Using one commonly used scenario for the extent of global heating by the end of this century, the prediction was for a substantial reduction in the extent of suitable climate conditions (although Britain – except for the southeast – was still within the zone of suitable climate). Although such modelling approaches, known as species distribution models, have been widely applied to larger organisms, as far as I am aware this study was the first time it had been attempted for a free-living protist.

A comparison between these two studies of the global distribution of *A. vas* also illustrates how approaches to this area of science have been developing over the last few decades. In the 2007 study Humphrey Smith and I compiled a global data set for *A. vas* distribution by searching large numbers of published research papers looking for studies that recorded the presence of the species.[17] Much of this very time-consuming work was carried out by Humphrey, as by that point he had retired from lecturing and had the time available for such a labour-intensive study, while I still had a high teaching load. By contrast, for the 2024 study Estelle Bruni worked with 27 co-authors (including myself in a minor role).[18] So this new global data set was constructed as a group effort, often by people who had either collected the data themselves, and/or were familiar with that particular part of the world and fluent in the various languages used in the many relevant past publications. This led to a more extensive, and more accurate, set of data than Humphrey and I had been able to compile between the two of us. In addition, several of the authors had particular expertise in the species distribution modelling aspects of the study. This multi-author approach, with its obvious advantages for compiling global data sets and bringing in collaborators with particular skill sets, has become much commoner over the last few decades.

Apodera vas also illustrates another important aspect of the microworld: the potential role of humans in moving microbes around the planet. This has long been studied in a marine context, where plankton can be transported in ships' ballast water. When empty, a large cargo ship becomes unstable, and a common solution is to fill ballast tanks with seawater, which is then released when the

a

b

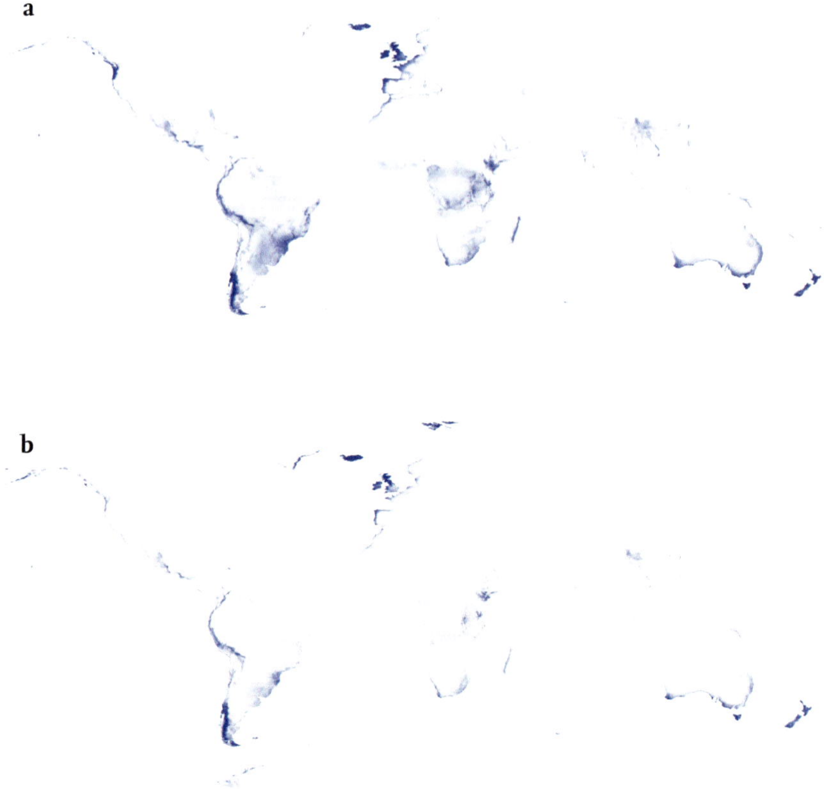

FIG 114. Modelled potential distribution of *Apodera vas* based on climate. Map (a) shows areas today with suitable climate – the darker the shading the more suitable the climate. This map is based on air temperature, but a model using soil temperatures gave a very similar pattern. Map (b) shows modelled distribution for one example of predicted climate 2071–2110 (other climate change scenarios gave broadly similar patterns, showing a contraction in suitable climate). (Redrawn from Bruni *et al.* 2024; Creative Commons licence, CC BY 4.0)

ship is ready to be loaded again. Large ships can take on over 30,000 tonnes of water, moving huge quantities of small marine organisms around the world.[19] There are similar issues in freshwater too, albeit usually on a smaller scale, with boats and other equipment potentially moving organisms between sites – although it's usually macroscopic organisms that cause people the most concern (Figure 115).

FIG 115. Warning sign at Blakeney in Norfolk alerting boat owners to the potential spread of 'invasive' aquatic species facilitated by moving boats between sites. In this case the emphasis is on macroscopic invertebrates and plants, but many freshwater microorganisms must be moved between river catchments in this way as well.

Until recently, the human movement of soil microbes has received very little attention, but we must have been moving soil around the planet for as long as long-distance seafaring has existed. When I became interested in this topic some 15 or more years ago I discovered that bog-moss *Sphagnum* was being imported to Britain from New Zealand for use in orchid cultivation and in hanging baskets. I bought some of this moss and found a large number of live microbes such as various testates, ciliates, bdelloid rotifers and nematodes – all imported from New Zealand. The commonest testate was *Argynnia dentistoma* (old name *Nebela dentistoma*), but the moss also contained examples of *A. vas* (which also used to be placed in the genus *Nebela*).[20] It would seem very likely that *A. vas* will now be living in the wild among the mosses in some British gardens, having escaped from hanging baskets, as we know from the work described above that the British climate seems very suitable for this species.

Following my short 2010 paper on this topic, Allison Perrigo and colleagues sampled dried mud from 18 pairs of boots used for fieldwork and found viable dictyostelid slime mould of several species from four of the boots, along with unidentified plasmodial slime moulds from seven pairs of boots.[21] I had previously found testate amoebae, ciliates and bdelloid rotifers in small quantities of soil washed from my walking boots on return to Britain from fieldwork in Canada – all viable despite intercontinental travel involving x-rays, and then several months in a laboratory freezer before I had time to examine the sample. The interesting question is how often this sort of human-assisted dispersal leads to a species colonising a new site. There are almost no studies looking at this for soil microbes, but one of the very few is from the Arctic.

THE POLAR STORY: INTRODUCED SPECIES AND CLIMATE CHANGE AT HIGH LATITUDE

An obvious way to investigate the potential for human introductions to affect the ecology of soil protists would be to introduce microbes from soils in one part of the world to new sites and see what happens. The problem is that this would be at best ethically dubious – what if you introduce something that becomes common and has significant negative effects on the local environment? However, sometimes history sets up an experiment for you. Pyramiden is a now abandoned settlement on the Arctic island of Spitsbergen, in the Svalbard archipelago (Figure 116). It was founded as a coal mining town by Sweden in 1910 before being sold to the USSR in 1927, and was finally abandoned in 1998. Sometime in the early 1980s large amounts of soil were imported to the site, apparently from the steppes in the southern part of the former USSR. The idea was to improve the

FIG 116. The former Russian mining settlement of Pyramiden on the Arctic island of Spitsbergen; it is now abandoned, except for some tourist visits during the summer. The study of testate amoebae described in the main text was carried out by Mazei *et al.* (2018), while the related study of soil invertebrates was by Coulson *et al.* (2013). These papers (especially the 2018 one) also detail the history of the site and what is known about the imported soil. (Steve Coulson)

local soil for growing food (especially in greenhouses), and to generally 'green' the rather bleak site. In addition, it seems that the soils were used as solid ballast in the ships coming to Pyramiden to pick up the coal, so were effectively imported at little cost. Soils from the steppes will likely have contained many organisms not native to an Arctic island, and they were sampled some 30 years later to see if this translocation of soil had led to the successful introduction of new species.

These soils were studied for testate amoebae, and compared with the testates in non-imported soils from nearby.[22] Several species of testate were found in the imported soils but not in the local soils. However, only one of these had never been recorded in Svalbard, but it didn't feature in testate samples from the steppes either. This contrasts with studies of soil invertebrates from the Pyramiden soils, which found 11 species never before found in the Svalbard archipelago (24 per cent of all invertebrate species recorded from the site in this study).[23] Most of these were small invertebrates – such as springtails (Collembola), and gamasid mites – which because of their small size sit at the interface between the micro- and macroworlds. Yuri Mazei and colleagues suggested that the lack of any obvious introduced testates was in part because they tend to have very wide distributions (while specifically noting that there are exceptions to this, such as *A. vas*). They concluded that although microbial 'species introductions due to human activity may well occur, our data suggests they may be rare'.[24] However, we currently have very few studies like this Pyramiden one from which to generalise.

One interesting thing about the mites and springtails from the steppes is that they have come from a climate that is much warmer than the Arctic in summer, although much of the steppes can be very cold in winter. However, they have been introduced to a location that is warming more rapidly than most of the planet (Figure 117). High latitudes contain some of the most rapidly warming areas, with potentially substantial effects on the microworld. Some 30 years ago, in one of the first papers to consider testate amoebae in the context of human-driven climate change, I suggested that testates might be very useful in monitoring the biological effects of global heating in the Antarctic.[25] More recent work has confirmed their usefulness.

The Antarctic Peninsula is relatively ice-free, and so has more wildlife than most of the continent. This is why it's a major destination for cruise ships containing tourists interested in natural history. As with Svalbard in the Arctic (a similar, if slightly cheaper, ecotourism destination), the Antarctic Peninsula is warming rapidly. One difficulty in looking for any effects of this warming in the microworld is that, for very obvious reasons, there is a shortage of past data to compare with what we see today. Even now, Antarctica is a tough and expensive place to do fieldwork; this was even more the case in the past. Testate amoebae

FIG 117. Svalbard: a view of the mountains near Pyramiden. Polar regions are warming faster than most of the rest of the planet. The Barents Sea, which includes the Svalbard archipelago, is warming especially fast – including very rapid reductions in the extent of sea ice (Descamps *et al.*, 2017). (Steve Coulson)

provide one way round this problem. Some of the deep cushions of moss can be shown to be at least 150 years old at their base by using radiocarbon dating and other methods. The shells of testate amoebae are preserved in these peaty moss banks and show increases in population sizes over the past 50 years as conditions have warmed – it's likely that both temperature itself and temperature-related availability of liquid water have been important.[26] Mossy, peaty habitats are also important habitats in Britain and Ireland, and testates, along with other protists, are increasingly playing a major role in our understanding of how bogs respond to a changing climate. However, before considering this it's worth taking a look at a recent study of the effects of expanding urban environments on soil protists.

A CHINESE STORY: URBAN LAND USE AND SOIL PROTISTS

As well as changing the climate, our activity has been changing the nature of habitats for a much longer time. Consider the view from the Selborne Hanger: all the habitats you see have been heavily modified by humans. The most obvious example is the village, but all the fields, hedgerows and woodlands are a product of many centuries of human use too. There is a long history of studies on how the changes we make to the landscape affect larger organisms such as birds, and as this chapter shows there is now an increasing number of studies asking similar questions about the microworld.

Selborne is a small village, but globally a very obvious land use that has profound effects on an area's natural history is the rise of large cities. China provides a spectacular example because of its recent urbanisation following its highly successful economic reforms in the late twentieth century. For most of my lifetime (I was born in the early 1960s) China was a predominantly rural country, but by 2013 over 50 per cent of the population were living in cities. This obviously has implications for the nature of Chinese habitats: for example, over the first 12 years of the twenty-first century the country lost more than 32,000 square kilometres of agricultural land to urbanisation.[27]

As a large country crossing several climate zones with many very large cities, China makes a good place to look at the effects of urbanisation on soil protists. To do this Zhi-Peng Li and colleagues used molecular methods (metabarcoding) to compare the urban microworld in a series of temperate and subtropical Chinese cities (Figure 118).[28] They found that, compared to nearby forests, the more urban habitats extended the distribution range of protists with broader niches – that is, microbes that could survive under a wider range of conditions. This is what we are used to seeing in larger organisms too. Consider birds: the adaptable sparrows have benefited from human habitats and are now widely distributed in cities throughout the world, although in China the Tree Sparrow *Passer montanus* takes on the urban role filled by the House Sparrow *P. domesticus* in Britain.[29] Birdwatchers usually have to leave urban areas for more 'natural' habitats if they want to see species with more specialised ecological requirements, but Tree Sparrows can be found happily foraging among the tourists around the Xi'an city walls shown in Figure 118.

The Chinese study also found that urbanisation particularly favoured the spread of temperate protists south into subtropical areas. They suggested that soil pH is key to understanding this. The subtropical soils were generally more acidic, but urbanisation tended to increase soil pH – that is, to make them less acidic, and so more suitable for some temperate-zone protists. Soil pH seems to be particularly important to bacteria, so one likely explanation is that the increased pH in urban areas is affecting many protists indirectly by increasing their bacterial prey. Northward expansion of subtropical soil protists was less common, and possibly restricted both by higher pH (a potential problem for the more acid-loving subtropical species), and also the drier conditions of central

FIG 118. Li and colleagues (2025) sampled soils from 13 cities across subtropical (mean annual temperatures > 15 °C) and temperate (4–13 °C) China. Around each city they sampled relatively 'natural' forest areas, along with more highly modified farmland, urban parks and residential areas. (a) One of the temperate cities used was Xi'an in central China; the photograph was taken on the spectacular historic city walls – with the old city to the left and the tower blocks of the much larger modern Xi'an catching the evening light. (b) One of the subtropical cities sampled was Xiamen on the southeast coast of China. The photograph was taken from the island of Gulangyu, with its European-style architecture reminiscent of the Mediterranean, looking towards the modern city in the pollution haze beyond. Gulangyu became a British trading base in the nineteenth century, following the First Opium War. Before clearance for agriculture this part of China was mainly tropical forest (Cunliffe, 2015). Much of my own work on freshwater microbial ecology has been on sites in the Xiamen area.

China. However, the key result that Zhi-Peng Li *et al.* chose to highlight was that 'human land use practices facilitate the range expansion of generalist taxa of soil protists', leading to a homogenisation of soil protists, with urban areas in different climate zones developing similar-looking species lists.[30] This is very like what we see in larger organisms, with sparrows, rats and mice being common urban species around the globe. Climate, pH and moisture content also play key roles in my next story of protists and global change in peatland habitats.

THE PEATLAND STORY: TESTATES AS KEY PLAYERS IN PEATLAND CONSERVATION

Peatlands play a potentially important role in climate change, as although they cover only around 3 per cent of total global land surface they contain around one-third of all soil carbon. These are often threatened habitats, and their preservation and restoration have become an important environmental issue.[31] In Britain and Ireland peatlands form an internationally important habitat, with around 7 per cent of the UK covered by peat to a depth of at least 30 cm, while the figure for Ireland is 13 per cent.[32] Concerns about climate change have caused attitudes to peatland conservation in Britain to change significantly over the last 50 years or so. In 1993 Richard Lindsay argued that peatlands had been something of a Cinderella habitat as far as British conservationists were concerned, but at last people were waking up to the importance of their wildlife and its conservation.[33]

Initially, the focus was on the conservation of peatland habitats and species, but even 30 years ago scientists were starting to become aware that peatlands were important in the global carbon cycle. Just a few years after Lindsay's article, the leading academic ecologist John Lawton wrote that there was 'a new resurgence of interest with the realisation that the world's peatlands contain roughly as much carbon as the world's forests' in an essay in a high-profile ecology journal.[34] He suggested that concerns about global warming could bring renewed interest, and research funding, to peatland ecology. Over the next few decades this prediction became true. Certainly, in my own university lectures on peatland ecology and conservation, there was a gradual increase in the emphasis I gave to peatlands as carbon stores through the 1990s and into the twenty-first century.

One effect of this increase in interest in peatland ecology and conservation can be seen in the growth of testate amoeba research in Britain. With a few exceptions, most British testate research in the last few decades has been focused on peatlands. This work has been carried out particularly in geography departments, using the shells of testates preserved in peat to reconstruct the past wetness of bogs. While all testates like reasonably wet conditions, some species like it wetter than others, allowing attempts to reconstruct changes in bog wetness over thousands of years by identifying changes in testate species in cores of peat extracted from bogs. In addition, there have been multiple studies of testates living on the surface of bogs, with the idea that this should tell us useful things about the current wetness, and so 'health', of the bog. This is potentially important in trying to assess the trajectory of peatland conservation schemes.

In the context of climate change, and the amount of carbon stored in peatlands, there has been a growing interest in restoring bogs that have been modified by humans. In the past many peatlands have been drained to make them more suitable for conventional agriculture, or to allow the peat to be extracted for use as fuel or in horticulture. Therefore the main approach to restoring such sites to functioning peatlands is to reverse the drainage process – often by blocking drainage ditches. Testates can be used as a way of investigating the success of such schemes (Figure 119). In a recent study in Northern Ireland some species of testates quickly responded to such drain blocking, even when direct measurements of changes in the water table were giving equivocal results.[35] However, there seems to be variation between sites, and two other recent British studies suggested that testates are often more likely to respond on a timescale of decades, while the vegetation responded much more rapidly to the restoration management.[36]

It seems clear that testate amoebae can be useful in monitoring the success of such restorations, although the associated microscope work can be very time-

FIG 119. Peat cutting on Chat Moss near Manchester (photographed in 2004). The peat was being sold for horticultural use. This now forms part of the area that is currently undergoing restoration with the aim of re-establishing a functioning peat bog. The restoration involves making the peat wetter again, which should lead to the process of peat formation restarting. The nearby Astley Moss (part of the wider Chat Moss complex) has been undergoing restoration since the 1980s, while the restoration started on the area shown in the photo in 2013 (Wilkinson, 2023; Osborne *et al.*, 2024). Testate amoebae can be used both to identify past conditions at the site and to describe the current conditions as a way of quantifying the success of restoration. For testates at the Astley Moss site see Davis & Wilkinson (2004); this was the first British study to use testate amoebae in the context of peatland restoration.

consuming. We need a greater understanding of the details of the responses to restoration if we are to make best use of testates in this context. In future, the labour-intensive microscopy work may possibly be replaced with molecular methods (eDNA) – as is now often the case in surveying ponds for amphibians.

A good way to try and understand the effects of a warming climate on the ecology of peat bogs is to experimentally warm the bog surface and see what happens. Experiments on a peatland in the French Jura have done this, warming small areas of bog surface using open-top chambers – which are a bit like mini greenhouses but lacking a roof, so air can circulate and the rain can get in (Figure 120). Within the chambers peak summer temperatures were 2–5 °C higher than on the surrounding open bog surface. Over a five-year period this warming was associated with a 50 per cent decrease in the biomass of mixotrophic testate amoebae. Many peatland testates, such as *Hyalosphenia papilio* (see Figure 42) are mixotrophic, able to make use of light energy because

FIG 120. The peat-bog warming experiment at Forbonnet in the French Jura. I took this photo in 2009, about a year after the start of the experiment. The walkway leads to six warming plots, and six control plots where no warming took place. Open-topped chambers were used to warm the bog surface – these are effectively mini-greenhouses but without a roof. The first plot seen on the left is a control plot (marked out for sampling, but with no open-topped chamber), while the second plot on the left is a warming plot. The board-walk allowed researchers to access the experiment without disturbing the bog surface. See Jassey *et al.* (2015) for the results of this warming on mixotrophic testate amoebae, which are summarised in the main text.

they contain symbiotic algae. Mixotrophic testates can make up 70 per cent of the total biomass of microbes in some peatland habitats, so are clearly potentially important. If these results are typical, then climate warming could lead to reductions in the rate at which peatlands lock up carbon, by reducing photosynthesis by mixotrophic testates.

These testates are not the only photosynthetic microbes to be found in peatlands. In the previous chapter I described how in the late nineteenth century Eugenius Warming wondered if cyanobacteria living in moss might turn out to be important. More recent research shows that indeed they are, along with other photosynthetic bacteria (some of the proteobacteria). Also important are a wide diversity of photosynthetic protists – especially green algae.

Although many people – such as Warming – had wondered about photosynthetic microbes in the mosses that often dominate peatland vegetation, until recently most had assumed they were unimportant, and that photosynthesis on peatlands was all down to the much larger multicellular plants. Much of the recent work in the area has been led by Vincent Jassey; his earlier work was often on testate amoebae in bogs (predators, albeit ones that often house symbiotic algae too), but he 'soon noticed the high abundance of photosynthetic organisms'. Reflecting on this work, he pointed out that 'these abundant microorganisms were largely overlooked in studies of peatland carbon cycling', but he started to realise that they might be important in the context of climate change.[37] Vincent's group experimentally moved small blocks of bog surface around northern Europe – to see what happened to the photosynthetic microbes when moved to a different climate. They found that in this case increasing temperature increased the rate of microbial photosynthesis.[38]

Another detailed study of photosynthetic microbes in northern European peatlands found on average 7.5×10^7 photosynthetic cells per gram of dried moss from the bog surface.[39] This study estimated that up to 10 per cent of all the carbon fixed by a peat bog might be due to photosynthesis by microbes rather than by larger plants. So these photosynthetic microorganisms may have a vital role to play in determining how peatlands respond to a changing climate.

What do these various results tell us about the likely effects of warming climates on the ability of peatlands to take in carbon dioxide from the air via photosynthesis, and store it, out of harm's way, as peat? The answer is that we are currently unsure; this is very much work in progress, and peatland ecology can be complex. However, it's possible to make some tentative suggestions.

Increased temperature is likely to speed up the decomposition of peat, as microbial action tends to increase with temperature. The results of the experimental movement of peat blocks around northern Europe suggest that warming may increase microbial photosynthesis on bogs, which may in part offset the effects of warming on peat decomposition. While peat is broken down more quickly it is also created at a faster rate too. However, the fact that warming can be bad for mixotrophic testate amoebae is an added complication (in passing, it's worth pointing out that mixotrophic testates appear to be slow in recolonising peatlands under restoration management, which has implications for their ability to make new peat).[40] With peatlands, by definition a soggy habitat, the effect of a changing climate on rainfall will often be at least as important as any direct effect of temperature. These various interacting effects may differ from bog to bog, but over the last decade or so it has become clear that photosynthetic microbes, such as cyanobacteria, green algae and mixotrophic testates, need to be taken seriously when trying to work out the effects of climate change on peatlands.

When looking at microbial samples from bog-moss under a microscope you often see diatoms too. On bogs these can be particularly important on areas where conditions have caused larger plants to die back, allowing more light to the peat surface.[41] However, diatoms have been particularly used in studies of water quality in lakes and rivers. These form the last of my series of stories illustrating the importance of microbial protists in the context of global change.

THE DIATOM STORY: THE IMPORTANCE OF PROTISTS IN ENVIRONMENTAL MONITORING

Diatoms are often an important component of the algae found in both rivers and lakes. The headwaters of the River Tees provide a good example: in a survey in the 1990s diatoms were the most species-rich algal group found growing on the rocks in a stream (Harwood Beck), followed by green algae and cyanobacteria (Figure 121).[42] Diatoms not only grow as biofilms on rocks, making them slippery under foot, but they can be planktonic too. In a fast-flowing river such as the headwaters of the Tees any diatoms living as free-floating plankton would be rapidly swept away, but in lakes and reservoirs this is less of a problem. Upper Teesdale has such a site at Cow Green Reservoir, which was created at the start of the 1970s. The reservoir was controversial because of its ecological impacts, and this resulted in funding being made available to investigate its environmental effects, including studying the algae that colonised it. This work found that diatoms also dominated the reservoir algae; initially they were the same species that can be found in the local streams, but later more planktonic species arrived, possibly introduced by birds.

FIG 121. The River Tees near Widdy Bank Farm in Upper Teesdale; just downstream from here Harwood Beck flows into the Tees; this beck was the location for the 1990s algal survey described in the text. A short walk upstream from here leads to Cow Green Reservoir. See Kelly *et al.* (2018) for a summary of the freshwater algae of Upper Teesdale, including Cow Green Reservoir and Harwood Beck.

Diatoms have been particularly important in monitoring water quality in rivers, since pioneering work by Ruth Patrick in North America in the mid-twentieth century.[43] Benthic (bottom-living) river algae, such as those in the Tees, became increasingly prominent in the late twentieth and early twenty-first centuries across most European countries, including Britain, as a routine part of the ecological assessment of river waters. One problem is that these diatom communities can be very species-rich, and because of this many approaches to diatom-based monitoring use genus-level data, or concentrate on a smaller number of indicator species thought to be sensitive to particular types of pollution. For example, the Trophic Diatom Index, first developed in the 1990s, used 86 species chosen for their sensitivity to nutrient pollution and ease of identification to score nutrient levels on a five-point scale.[44] Diatoms can also be used as indicators of water acidity, and in this role played an important part in establishing the connection between acid rain and the acidity of lake waters.

When I was a student in the 1980s acid rain, and the increasingly acidic nature of many lakes and rivers, was a major environmental issue. Burning coal, especially in large power stations, created acid rain that was being blamed for the destruction of forests and the acidification of lakes – with associated loss of fish and other organisms. Before the mid-twentieth century, much of this pollution was relatively local, but with power stations acquiring taller smokestacks (to reduce local pollution) the pollution became long-range, affecting places far from the site of coal burning.[45] However, in the early 1980s both the British government and the Central Electricity Generating Board (the state power company at that time) were inclined to blame the acidification on changes to the vegetation surrounding the lakes, due to an increase in upland forestry. If this were true, the coal-fired power stations would conveniently be blameless. However, diatoms provide key evidence showing that air pollution was heavily involved in the acidification of British lakes.[46]

Diatom frustules preserve well in lake sediments, and this allows them to be used to reconstruct changes over time. The diatoms sampled from different depths down a core of lake sediment can be extracted, identified and counted. Different diatom species have different preferences in relation to water acidity – for example, *Pantocsekiella comensis* (formerly *Cyclotella comensis*) declines as waters become more acidic. Pollen grains also preserve in these sediment cores, and these can be used to reconstruct changes in the surrounding vegetation – such as afforestation of the catchment with conifers during the twentieth century. These studies, especially of several lochs in Galloway in southwest Scotland, showed that acidification started *before* afforestation. However, increased acidity did tend to be associated with markers of the Industrial Revolution, such as soot particles in the sediments. As the amount of coal we were burning went up, the pH of many lakes declined. This demonstrated that air pollution was the main driver of the acidification, although afforestation could make the situation worse, as air pollution can be deposited on the leaf surfaces and then washed into waterbodies. There are many more problems with coal burning, in addition to its well-known effect of increasing the concentration of carbon dioxide in the atmosphere.

MICROWORLDS AND GLOBAL CHANGE

As the various examples described in this chapter show, the microworld can be very useful in monitoring the environment. Diatoms can be used to assess water pollution, and testate amoebae the wetness of bog surfaces. However, even more important is the dominant role microorganisms play in most key ecological processes, as discussed in the previous chapter. In this chapter, I have used the example of carbon storage in peat bogs and its interaction with human-caused climate change; however, as Philip Donoghue and his co-authors recently emphasised, 'The Earth system today is driven by an interlinked network of biogeochemical cycles that are governed largely by microbes'.[47] Some 20 years ago Tom Curtis made the same point in a rather more dramatic way, writing that 'if the last blue whale choked on the last panda, it would be disastrous but not the end of the world.' However, if the last two microbial species responsible for a key link in the nitrogen cycle became extinct, 'that would be another matter. It could be happening now and we wouldn't even know'.[48] The philosopher Maureen O'Malley – who is fond of this quote – suggests that it 'captures very nicely the tensions in conservation biology: who or what is conservation for? Us, or other biological entities? Nature in all its biodiversity for its own sake?'.[49]

You can ask the question at an even larger scale too. Conservation for the sake of the whole planet? The actions of the microworld affect not only biology, but the physics and chemistry of the planet – and, as the examples in this book show, this ranges from oxygen from microbial photosynthesis, to the cycling of carbon, to the structure of soils.

The microworld needs to be taken much more seriously when we think about conservation and the environment. Without the ecological roles of microbes the view from the Selborne Hanger would be of a bare and lifeless planet. Up to now, much research on the role of microbes in ecological processes has focused on prokaryotes and fungi. For example there was a big increase in the number of studies on bacteria and fungi in soils during the 1990s and the first decade of the twenty-first century, while the number of protist studies largely remained steady.[50] Things are starting to change, however, and as this chapter has shown there is today an increasing interest in the role of protists in a wide range of environmental issues. One important aim of this book has been to help raise the profile of algae and protozoa – both for their intrinsic interest and beauty, and also for the major contribution they make to the functioning of our planet.

APPENDIX

A Brief Guide to Finding Protists

This short appendix provides suggestions for finding a range of protists for anyone with access to at least a basic microscope, ideally with the ability to magnify to at least ×200, although lower-power stereomicroscopes (×10–40) can be of use for slime moulds and some other larger members of the microworld. For the most part, I assume you have no access to laboratory equipment beyond the sorts of things that can be found in a kitchen, or sometimes basic equipment that can be cheaply bought from a specialist natural history equipment website. Many of the organisms featured in this book can be found in a wide range of habitats, but mosses are often a good place to start (soils are more difficult, as all the soil particles can obscure the organisms on your microscope slide). I first describe simple approaches to finding protists (especially testate amoebae) in mosses, before moving on to ways of finding algae, and also simple methods of culturing protozoa in general, and plasmodial slime moulds in particular.

Identification can be difficult, and many microbial groups do not have good general up-to-date identification guides. Brian Moss's book on *Ponds and Small Lakes* (2017) provides simple identification keys for freshwater microbes – both algae and protozoa – while Bellinger & Sigee's guide to freshwater algae (2015) includes accessible keys (and many illustrations) to allow these organisms to be identified to genus level. Although a quarter of a century old, Martyn Kelly's key to the genera of benthic diatoms in streams and rivers (2000) is still useful, and designed to be used by non-specialists. There are no accessible up-to-date identification guides for testate amoebae – with the exception of the hyalospheniid testates, covered by Anush Kosakyan *et al.* (2025) – this monograph is freely available to download as a PDF, but I find the hard-copy version easier to use when sitting at a microscope. However, as described in the Prologue, Sally Corbet's 'Illustrated introduction to the testate rhizopods in *Sphagnum*' (1973) is a simple key to testate amoebae in British bog-mosses – while the names are now often out of date, it is still a useful starting point as it was designed for beginners, and has the advantage of being in *Field Studies*, a journal that is free to download from the Field Studies Council's website (www.field-studies-council.org). Bruce Ing's *Myxomycetes of Britain and Ireland* (2020) is a key to plasmodial slime mould species, but requires some experience of the group to use for reliable identifications. There are also many useful websites to help with identifications – although these can be less permanent than hard-copy publications. At the time of writing a very useful source of information on the identification of amoebae (both testate and naked) is Ferry Siemensma's *Microworld* site (https://arcella.nl); some of his fine amoeba photographs can also be seen illustrating this book.

One important aspect of the microworld is that you don't have to travel far to find unstudied sites. One example of my own is a study of the testate amoebae and diatoms in a nature reserve built round an old sand extraction site in Lancashire (Wilkinson *et al.*, 2017) – before this work no species in either group had been recorded from the reserve. The testate amoebae living in urban mosses are another accessible group – Mitchell *et al.* (2013) found 23 species in a small number of moss samples from Neuchâtel Castle in Switzerland. Microscopic animals can be similarly diverse and in need of study, as shown by an investigation of tardigrades from a local nature reserve in York which found three species new to the county (Thomas & Buckton, 2023). All these studies only used microscopy (i.e. no molecular methods requiring significant laboratory facilities), so are well within the capacity of a home or school lab. As Edward Mitchell and colleagues commented in their study of Neuchâtel Castle, their results illustrated 'the potential for discovery of a poorly known diversity within the microbial world, including in regions where the diversity of living organisms has been well studied for centuries'.

SAMPLING MOSS FOR TESTATE AMOEBAE: A SIMPLE WAY INTO THE MICROWORLD

Dense patches of damp moss are one of the easiest habitats to sample. Bog-moss *Sphagnum* can be particularly rich in protists and microscopic animals, and Marjorie Hingley's *Microscopic Life in Sphagnum* (1993) is a useful basic guide, albeit now rather out of date. However, all mosses will contain protists, and any moss present in quantity should provide an interesting selection of organisms, especially testate amoebae (Figure 122). With very wet moss, such as some bog-moss samples, the simplest approach is to squeeze the moss and then put a small amount of the resulting water on a slide under a microscope (if you have one available, a plastic disposable pipette is very useful for this, but you can often just squeeze a small amount of water directly onto the microscope slide). The larger testate amoebae should be apparent at a low magnification, but a magnification of around ×400 will be better for seeing details and smaller species (Figure 123).

For drier mosses, a better approach is to shake them in a sample tube with some water and then put a few drops of that water onto a microscope slide. A somewhat more sophisticated approach is to vigorously shake your moss sample with water to dislodge the testates, then sieve out the moss fragments (a tea strainer or similar will do, but if you have access to laboratory sieves use 250–300 μm). With access to lab facilities the sieved liquid would often be

FIG 122. Little Shaggy-moss *Rhytidiadelphus loreus* covering boulders and a drystone wall on the side of Cadair Idris in north Wales. Thick patches of this moss can contain a range of testate amoebae, including some of the larger and more distinctive species that are often associated with bog-moss (such as *Nebela collaris* and *Assulina muscorum* in examples of this moss I have looked at from the English Lake District).

FIG 123. An unusually diverse slide showing the microworld in bog-moss *Sphagnum* – including many testate amoebae, along with algae, pollen grains and rotifers. Among the testate amoebae are *Galeripora artocrea* (the large disc just left of the centre), *Hyalosphenia papilio* (the large yellowish test towards the top), *H. elegans* (breaking the frame in the top right), *Assulina seminulum* (only partly in shot in the top left-hand corner), along with two different species of *Euglypha*. Testates such as *Galeripora* (and the similar-looking *Arcella*) and *H. papilio* should be easy to spot at relatively low magnification (even with just a low-power stereomicroscope). (Edward Mitchell)

spun in a centrifuge (at around 3,000 rpm for 3–5 minutes) to concentrate the testates and other organisms at the bottom of the tube. However, without such facilities (including at remote field sites) the tube can be left to stand for a day or so, somewhere away from disturbance and vibrations, and gravity will tend to concentrate the testates at the bottom of the tube. You can then pour off the overlying water and pipette this residue onto your microscope slide. Mosses, especially species that form compact cushions, are also a good habitat for finding tardigrades.

HUNTING ALGAE WITH A TOOTHBRUSH

Algae are found in lakes, rivers, streams and ponds – on surfaces such as rock, mud and aquatic plants, and also in the waterbody. The ones in the waterbody are

FIG 124. The River Derwent in Borrowdale in the English Lake District. Scrubbing some of the small stones (as seen in the foreground of the photo) from the riverbed with a toothbrush produced a range of algae – including lots of filamentous green algae (e.g. *Mougeotia*), along with diatoms (e.g. *Nitzschia* and *Tabellaria*) and desmids (such as *Xanthidium*).

often sampled using a plankton net (as seen in Figure 107), but nets with a filter of a size suitable for algae are quite expensive (nets for sampling slightly larger zooplankton are somewhat cheaper). Algae scraped from surfaces are easier to sample. A good simple approach is to sample the biofilm on rocks and stones in lakes and rivers (Figure 124). This can be done using a toothbrush to scrape off the biofilm into some water in a small dish or tray. This water can then be put in a sample tube and taken home to look at under a microscope. As with the water from moss samples described above, if it is left to stand gravity will help concentrate the algae (and often some protozoa too) at the bottom of the sample tube. A more detailed account of sampling aquatic microbes with minimal equipment is given by Moss (2017).

Desmids – particularly beautiful green algae – are best searched for in acidic pools on peatlands and in the uplands. Examining bits of pond weed with a hand lens, or stereomicroscope, can sometimes find near-microscopic animals such as *Hydra*.

A SIMPLE APPROACH TO CULTURING PROTOZOA FROM SOILS AND FRESHWATER

In their *Introduction to the Study of Protozoa* (1961), Mackinnon & Hawes suggested the following simple approach to culturing protozoa for microscopic analysis just using grains of wheat or rice. They recommended a large pond or slow-flowing stream with lots of vegetation as habitats that are likely to provide good quantities of protozoa for study. As described in the discussion of Priest Pot in Chapter 5, many of the protists may not be active but only present as resting forms (cysts). Culturing (effectively improving the environmental conditions, often by increasing the food supply) can both cause protozoan cysts to become active, and can also increase the numbers of the other already active species, making them easier to find.

Many complex culture conditions can be used, but the simplest is to place boiled wheat or rice grains (brown rice is often recommended, as it is likely to contain more nutrients than white rice) in your sampled pond water – add 2–4 grains per 100 cm^3. The idea is that this provides food for bacteria, which are themselves the food source for many types of protozoa. Ideally, keep the cultures cool and away from strong sunshine, and examine them daily. This is effectively a slightly more controlled version of the hay-infusion approach described in Chapter 1.

CULTURING SLIME MOULDS

Plasmodial slime moulds are very common in the environment, but we only notice them when they produce visible fruiting bodies. These can be encouraged to form from bark or leaf-litter samples using moist chamber cultures (as described by Feest, 1996, or Ing, 2020). Place the plant samples in a Petri dish with some filter paper covering its base to retain moisture (alternatively any small shallow dish, with a cover, will do, and kitchen towel can be used instead of filter paper). Add pure water (without access to pure water from a laboratory still, the best approach is to use still mineral water) to flood the dish for 24 hours, then remove excess water (a pipette is useful for this) and keep covered and moist.

Over the next few days slime mould fruiting bodies may start to appear. Other species may take longer to become visible: for example, the photograph shown in Figure 125 was taken 12 days after the start of the culture (this was a leaf-litter sample, which would typically produce quicker results than a bark sample). New

FIG 125. Fruiting bodies of plasmodial slime moulds on leaf litter in a moist chamber culture. The photograph illustrates what you can do with basic facilities – in this case an old low-power (×8.75) stereomicroscope with a phone camera just pointed down the eyepiece. Better results can be achieved if a camera phone adapter is used (these are the same adapters that bird people use to take phone camera pictures with a telescope). See Figure 50 for an example of a photomicrograph made using a phone with such an adapter. The other photomicrographs in this book (except for Figure 2) were made using more sophisticated equipment.

species can still be appearing several months after the start of the culture. As a rule, if lots of fungi start to grow in your culture it's probably too dry, and if it starts to smell bad it's likely too wet, and bacteria are taking over (Ing, 2020). The slime mould fruiting bodies are of a size where a high-powered compound microscope is not required (except for details such as spores that, as with fungi, may be required to confirm identifications), and a hand lens or stereomicroscope can be used.

Glossary

Most of the terms included here are explained in more detail in the main text. The glossary is intended to act as an *aide memoire* for any unfamiliar terms that are used in multiple chapters, to remind the reader of their meaning. The entries aim to give the general idea behind the term, rather than a rigorous formal definition. In a few places I have taken the opportunity to include some additional detail that did not fit in the main narrative. Words in *italics* also have entries in the glossary.

This glossary is based on multiple sources, especially the glossaries in two of my previous books (Sherratt & Wilkinson, 2009; Wilkinson, 2023), along with Boenigk *et al.* (2015). This latter book, *Biodiversity and Earth History*, includes very useful summaries of all of the main higher taxa of eukaryotic organisms.

Alga (plural **Algae**) – These are *photosynthetic* organisms (often but not exclusively microbial) that are not members of the Embryophyta (the group of plants that comprises all the familiar types such as mosses, ferns, conifers and flowering plants). The term is mainly applied to *eukaryotes*, but the *cyanobacteria* are often included as ecologically similar to the eukaryotic algae. See 'The nature of algae' in Chapter 4, page 75.

Amoeba (plural **Amoebae**) – An informal term for a *polyphyletic* grouping of *heterotrophic* (or sometimes *mixotrophic*) microbes. Often split into naked (no shell) and testate (shelled) amoebae.

Amoebozoa – A *monophyletic* grouping of some (not all) *amoeba*-like organisms. See Figure 10, page 23.

Archaea – Originally called 'archaeobacteria'. As described in Chapter 1, towards the end of the 1970s Carl Woese and George Fox showed that 'bacteria' comprised two very different groups of organisms – the archaea and the 'true' *bacteria*, which are now usually considered to form two different *domains* of life.

Bacterium (plural **Bacteria**) – One of the two basic types of *prokaryotes*; the other is the *archaea*.

Bdelloid rotifer – See *rotifer*.

Benthic – Living on the bottom of a waterbody.

Biofilm – A film of microbial cells embedded in slime (more technically, embedded in extracellular polymers secreted by the microbes). For example, biofilms can be found covering stones in rivers and lakes.

Brown algae – A group of *algae* that are rare in freshwaters (see Table 2, page 76). Commoner in marine systems, where many common rocky-shore seaweeds are brown algae.

Brownian motion (**movement**) – Random motion of very small particles caused by the impact of invisible molecules.

Chloroplast – A structure within *eukaryotic* cells that contains the photosynthetic pigments. As described on page 37, they evolved from free-living photosynthetic bacteria (*cyanobacteria*). The term chloroplast was coined by Andreas Schimper in 1883: he also speculated that these chloroplasts may have had a *symbiotic* origin. However, he didn't develop these ideas further, but went on to be a key figure in the early history of plant ecology before dying at the age of only 45.

Chromosome – In *eukaryotes*, a microscopic structure within the nucleus holding genetic information arranged in a linear sequence. In *prokaryotes*, it is a circular structure not localised in a nucleus.

Ciliate – Informal name for the Ciliophora, a *monophyletic* group (*phylum*) of protozoa. See page 92.

Cilium (plural **Cilia**) – Very fine threads ('hairs') projecting from the cell surface. They function either to move the cell through water, or to move water past the cell to bring food particles within reach.

Cryptic species – Species which appear identical to each other using morphological features, but which can be separated by molecular methods. See Box 3, page 113.

Cyanobacteria – Oxygen-producing *photosynthetic* bacteria, often called blue-green *bacteria* (or blue-green *algae*) in the past. See Table 2, page 76.

Desmid – A type of *green alga*, mainly associated with acid waters, and often very beautiful when viewed with a microscope.

Diatom – Aquatic single-celled *eukaryotes* with silica-rich shells (*frustules*). See Table 2, page 76.

Dinoflagellate – A group of *planktonic protists*, which includes both *photosynthetic* and *heterotrophic* species. They move with the aid of two differently sized *flagella*. See Table 2, page 76.

Domain – A higher level of taxonomic classification than the more familiar *kingdom.*
Ediacaran – The name for both a geological period (from 635 to 538.8 million years ago), and an enigmatic early group of multicellular organisms that lived at that time. See the start of Chapter 7.
eDNA – Short for environmental DNA.
Encystment – The production of a resistant stage (cyst). This can allow the microbe to survive unfavourable conditions in a dormant state.
Endosymbiosis – One organism (e.g. an *alga*) living inside another (e.g. a *protozoan*).
Eukaryote (**Eukarya**) – One of the two great structural divisions of life on Earth; the other is the *prokaryotes.* Eukaryotes have a more complicated cell structure than prokaryotes, containing *organelles* such as *mitochondria.*
Flagellate – A *polyphyletic* informal grouping of *protozoa* with *flagella.* In the past, flagellates were often assumed to all be closely related to one another, but we now know this is not correct.
Flagellum (plural **flagella**) – In *eukaryotes* these are structurally similar to *cilia,* but are longer and fewer in number. *Bacterial* flagella have a different structure, although they look similar to eukaryotic flagella. See Figure 31, page 64.
Frustule – The silica-rich cell wall of a *diatom.*
Gaia – The idea, originally suggested by James Lovelock in the 1960s, that organisms and the physics and chemistry of the environment evolve as a single system, from which emerges the sustained self-regulation of climate and chemistry at a habitable state for whatever is current biota (i.e. life on the planet).
Genome – The total genetic material in a cell.
Green algae – An algal group which, unlike many *algae,* are still considered to be true plants. See Table 2, page 76.
Heterotrophy – The use of organic carbon (rather than *photosynthesis* or other non-organic sources) to provide energy.
Kingdom – Traditionally the highest level used in classification. In a two-kingdom classification the kingdoms were plants and animals. See Box 1, page 16.
LECA – The last eukaryotic common ancestor (see *molecular clock*).
LUCA – The last universal common ancestor (see *molecular clock*).
Meiosis – A type of cell division during which the number of sets of *chromosomes* is halved. Importantly, it typically involves genetic recombination by crossing-over (switching genetic material between chromosomes) as well as halving the total genetic material. In humans it is the process by which eggs and sperm come to contain half the amount of genetic material found in standard body cells.
Mitochondrion (plural **Mitochondria**) – An *organelle* in a *eukaryotic* cell which is the location for energy production.
Mixotrophy – Obtaining energy by both *heterotrophy* and *photosynthesis.*

Molecular clock – An approach to using the rate of change in biological molecules (e.g. genetic material or proteins) to estimate the time since the organisms which contain them last had a common ancestor. For example, estimating the age of *LECA*, the last common ancestor of all eukaryotes, or *LUCA*, the last common ancestor of all life on Earth.

Monophyletic – A group of individuals descended from a common ancestor. In contrast, a polyphyletic group does not have a single common ancestor.

Nematode – 'Roundworms', near-microscopic animals in the phylum Nematoda.

Nucleus – An *organelle* in a eukaryotic cell which contains the *chromosomes*, and hence most of the genetic material.

Organelle – Any structurally distinct part of a *eukaryotic* cell with a particular function (e.g. *mitochondrion* or *nucleus*).

Phagocytosis – The engulfing of solid particles by a *eukaryotic* cell by the budding off of a vesicle from the cell membrane.

Phagotrophy – Feeding by *phagocytosis.*

Photosynthesis – The complex processes by which some organisms use light energy to produce organic chemicals from inorganic chemicals. Usually associated with plants, but many other organisms can also be photosynthetic.

Phylum (plural **phyla**) – A high level in the conventional classification of life just below *Kingdom.* For example in the animals (a *kingdom*) the *tardigrades, rotifers* and *nematodes* are all phyla. See Box 1, page 16.

Plankton – Organisms living in open water rather than on the bottom of a waterbody (see *benthic*).

Plastid – An *organelle* in the cell of a photosynthetic *eukaryote*, usually associated with *photosynthesis*, but plastids can sometimes have other functions such as energy storage. So the term covers a wider range of structures than *chloroplast.*

Polyphyletic – See *monophyletic.*

Precambrian – An informal term for the geological time prior to the Cambrian (which is marked by non-microscopic fossils becoming common). It runs from the formation of the Earth (a little over 4.5 billion years ago) until 538.8 million years ago – so it covers most of the history of the planet. For almost all of this time all life was microbial.

Prokaryote – An organism (*bacteria* or *archaea*) with its genetic material not localised in the *nucleus.* One of the two great structural divisions of life – see also *eukaryote.*

Protist – *Eukaryotic* organisms which are not plants, animals or fungi. Most of the organisms described in this book (e.g. *protozoa* and *eukaryotic algae*) are protists. The group is not *monophyletic.*

Protozoan – In the past often used as a term of classification, but now normally used informally to describe *heterotrophic* (also many *mixotrophic*) *eukaryotic* microbes. This is how it is used in this book.

Pseudopodia – A temporary projection from the cell surface of amoeba-like cells used for movement and/or ingesting nutrients. See Figures 42 (page 80) and 47 (page 91).

Red algae – A *phylum* of *algae*, very closely related to true plants. See Table 2, page 76.

Rhizopod – An older term for *amoeba*.

Rotifer – A near-microscopic *phylum* (the Rotifera) of aquatic animals. Bdelloid rotifers are a particularly interesting group of rotifers as all species are asexual, and the group appears to have abandoned *sex* some 80 million years ago.

Sex – In everyday English usually a synonym for copulation, but here used in a more technical sense. Sex is a process of genetic reorganisation in *eukaryotes* via *meiosis* and gamete fusion (in mammals this is the fusion of egg and sperm), and usually associated with the formation of new individuals genetically distinct from their parents. So sex may, or may not, involve copulation. Some protists have many different mating types – equivalent to multiple different 'sexes'.

Slime mould (or **slime mold**) – A term for several groups of *amoebae* with complex life cycles, involving multicellular stages (or in some cases giant cells). Slime moulds used to be considered fungi, hence 'moulds' in their name. Discussed in detail in Chapter 7.

Stoneworts – A group of multicellular green algae. See Figures 38 and 39 (pages 73 and 74).

Supergroup – An alternative high-level classification of *eukaryotic* organisms, based on molecular data. See Figure 10, page 23.

Symbiosis – In its original usage this describes two organisms living very closely together – so closely that they usually appear to be a single organism (e.g. *algae* within the cell of a *protozoan*, or the multiple organisms that form a lichen). This can include parasitic relationships as well as beneficial ones, and is the way the term is used in this book. However, the term is sometime restricted to describing only beneficial relationships (which may not involve close physical relationships, such as bees pollinating flowers).

Tardigrade – A *phylum* (the Tardigrada) of near-microscopic animals, sometimes called water bears.

Taxon (plural **taxa**) – Any group in a biological classification. Species, genus, *phylum* and *kingdom* are all taxa.

Testate amoebae (or **testates**) – A *polyphyletic* grouping of shell-building *protozoa* which are often common in soils and freshwater, especially in mosses, and in sediments of high organic matter content.

Endnotes

PROLOGUE

1 Pigott & Pigott, 1959.
2 Corbet, 1973.
3 Hooke, 1665.
4 Allen, 2010.
5 Ralfs, 1848, p. 1.
6 Bellinger & Sigee, 2015.

CHAPTER 1

1 My historical reconstruction is based on Jardine (2003), supplemented by Birch (1756) and Kington (2010).
2 Kemp, 2006.
3 Wooton, 2015.
4 Kemp, 2006, p. 181.
5 Jardine, 2003.
6 Lawson, 2016.
7 Gest, 2004.
8 Jardine, 2008.
9 Corliss, 2002.
10 Gest, 2004.
11 Ford, 2007a.
12 van Leeuwenhoek, 1674.
13 Corliss, 2002.
14 Corliss, 2002; Lane, 2015.
15 Gest, 2004.
16 Lane, 2015.
17 Ford, 2007a, 2007b.
18 Gest, 2004.
19 Gest, 2004.
20 Fara, 2009.
21 O'Malley, 2014.
22 Esteban & Fenchel, 2020.
23 Bonner, 2006, p. 45.
24 Mayr, 1982.
25 The 'he stands out as …' quotation comes from Midgley (1995, footnote 11 on page 260); italics as in her original text. The suggestion that she was 'steeped in Aristotle' was made to me by her son David in a conversation in 2024.
26 Aristotle, 1941, p. 635.
27 Warming, 1890.
28 Rothschild, 1989; Sapp, 2009.
29 Mann & Vanormelingen, 2013.
30 Cavalier-Smith, 2010.
31 For example, Cavalier-Smith, 2010; Ruggiero *et al.*, 2015; Esteban & Fenchel, 2020.
32 Margulis & Chapman, 2009.
33 Sapp, 2009.
34 Mackinnon & Hawes, 1961, p. 1.
35 Keeton, 1980.
36 Hagen, 2012; Wilkinson, 2021.
37 The 'child' has been dropped, and the pub is now known as 'The Eagle'.
38 Stent, 1981.
39 Olby, 2009.
40 Bronowski, 1968. Much later in his life, Watson told several people that the scene in the pub,

along with some other scenes, was largely invented for dramatic effect (Cobb, 2025).

41 Vickerman & Sleigh, 2000, p. 107.

42 The *Tomorrow's World* story came from Colin Curds, who worked at the Natural History Museum, in an email to me in July 2015.

43 For example, see Ogden & Hedley, 1980.

44 Katscher, 2004.

45 Translation from the French by Katscher, 2004, p. 260.

46 Keeton, 1980, p. 99.

47 Sapp, 2009.

48 Sapp, 2009; Pace *et al.*, 2012.

49 Pace *et al.*, 2012.

50 Eme & Doolittle, 2015.

51 For example, see Bhattacharya *et al.*, 1995; Wanner *et al.*, 1997.

52 Kosakyan *et al.*, 2016.

53 Clark *et al.*, 2018.

54 Useros *et al.*, 2024.

55 Wilson, 1994.

CHAPTER 2

1 Murray, 1987.

2 Grube & Balestrini, 2024, p. 1890.

3 Friend, 2012.

4 Lane & Xavier, 2024.

5 Much of this brief discussion is based on the more technical summary in Wilkinson, 2023, pp. 8–9.

6 Brasier *et al.*, 2002.

7 Westall, 2025.

8 Lane & Xavier, 2024.

9 Antcliffe *et al.*, 2017.

10 Porter & Knoll, 2000.

11 Porter *et al.*, 2003.

12 Lahr *et al.*, 2019.

13 Porfirio-Sousa *et al.*, 2024.

14 Porter & Knoll, 2000.

15 Porter, 2016.

16 Porter, 2016.

17 Wilkinson, 2023.

18 Sagan, 1967.

19 Margulis, 1971.

20 Margulis, 1998, p. 29.

21 Lane, 2017.

22 For instance, see Maynard Smith & Szathmáry, 1995.

23 Zachar *et al.*, 2018.

24 Martin *et al.*, 2017.

25 Lane, 2017.

26 Brasier, 2009.

27 Kirkland *et al.*, 2025.

28 A key source for my account of the Torridonian fossils and their environmental context is Brasier *et al.* (2017a), a book chapter giving an overview of this work with Alexander Brasier as the lead author. It had originally been intended to be a joint publication with his father. The Geological Society volume it appeared in was 'a celebration of the work of Martin Brasier'.

29 Brasier *et al.*, 2017b.

30 Mitchell *et al.*, 2021.

31 Lenton & Daines, 2017.

32 Strullu-Derrien *et al.*, 2019a.

33 Beerling, 2019.

34 Bower, 1935, p. 112.

35 Bower, 1935, p. 116.

36 Liston & Sanders, 2005.

37 Strullu-Derrien *et al.*, 2019a.

38 Wilkinson, 2021.

39 Sandon, 1927, p. 1.

40 Nielsen, 2019.

41 Darwin, 1839, p. 587.

42 Wilkinson & Smith, 2006.

43 Foissner *et al.*, 2002.

44 Christenhusz *et al.*, 2018.

45 Wilkinson & Smith, 2006.

46 Belnap, 2003; Mitchell *et al.*, 2021.

47 Bamforth, 2004.

48 Wilkinson *et al.*, 2025.

49 Fry *et al.*, 2024.

50 Wilson, 1994, p. 364.

51 Darwin, 1839, p. 394.

52 Segawa *et al.*, 2018; Procházková *et al.*, 2019.

53 Light & Belcher, 1968.

54 Cameron *et al.*, 2024.

55 Clarke, 2017.

56 Vincent, 1988.

57 Thompson *et al.*, 2019.

CHAPTER 3

1 Maynard Smith, 1968.
2 For example, see Ruxton, 2019.
3 Vogel, 2013, p. 553.
4 Beerling, 2019, p. 96.
5 Edwards & Kenrick, 2015.
6 Bonner, 2006, p. 39.
7 Haldane, 1927, p. 19. For a recent biography of J. B. S. Haldane, including a description of his popular science essays, see Subramanian (2020).
8 Vogel, 2013.
9 Vogel, 2013.
10 For example, see Monteith & Unsworth, 2008.
11 Wilkinson, 2001.
12 A good somewhat more technical description of viscosity – aimed at biologists – is given by Vogel (2013), and this is one of the key sources I used in writing this section. University-level physics textbooks will provide more detailed and rigorous discussion. However, as one of the most famous undergraduate physics textbooks explained, in the context of describing viscosity, 'Although you will see some complicated equations and hear about some complicated things, it's not our purpose that you should learn all these things' (Feynman *et al.*, 1964, p. 41).
13 Bonner, 2006.
14 Vogel, 2013.
15 Thorne & Blandford, 2017.
16 Wilkinson, 2005.
17 In the interests of historical accuracy, it's worth pointing out that the publication history of Brown's work is a bit more complex and ambiguous than my brief summary suggests. When Brown circulated it as a preprint (the contemporary description was 'printed for private circulation'), he also gave permission for the *Philosophical Magazine* to publish a copy of his first pamphlet in September 1828. However, despite this, the Ray Society's posthumous collection of Brown's work described it as a previously 'unpublished' manuscript.
18 Brown, 1866.
19 Merton, 1968.
20 Dusenbery, 2009.
21 Feynman *et al.*, 1964.
22 Dusenbery, 2009.
23 Andrews, 2017.

CHAPTER 4

1 Nicholson, 1957.
2 Allen, 1971.
3 Stancheva *et al.*, 2014.
4 Silva *et al.*, 1972.
5 Duigan *et al.*, 1998.
6 McGovern *et al.*, 2014.
7 West, 1911.
8 Duigan *et al.*, 1998.
9 Bolton, 2016.
10 Oxygenic photosynthesis is the photosynthesis we learn about at school, where oxygen is given off as a waste product (there are other types of photosynthesis used by some prokaryotes in which oxygen is not a waste product). Bolton even suggested using the term 'plants' for anything that conducts oxygenic photosynthesis. I am less convinced by this because of the confusion with the now normal taxonomic use of 'plants'. Sanders (2021) coined the term 'photoaerogens' for photosynthetic oxygen-producing organisms. However, the phrase 'plants and algae' seems the simplest solution for an appropriate non-taxonomic term.
11 Guiry, 2024.
12 Nave, 1867.
13 John *et al.*, 2011.
14 John *et al.*, 2022.
15 John *et al.*, 2011.
16 Guiry, 2024.
17 Priddle & Happey-Wood, 1983.
18 Wilkinson, 2021.
19 Barreno *et al.*, 2022.
20 Translation from Melkonian & Mollenhauer, 2005.
21 Melkonian & Mollenhauer, 2005.
22 Nowack, 2014.
23 Nomura *et al.*, 2014; Nomura & Ishida, 2016.
24 Nowack, 2014.
25 Marin *et al.*, 2005.
26 Nowack, 2014; Macorano & Nowack, 2021.
27 Van Etten *et al.*, 2023.

28 Gabr *et al.*, 2020.
29 Money, 2014; Sanders, 2021.
30 Mann & Vanormelingen, 2013.
31 Nave, 1867, p. 11.
32 Money, 2014.
33 Wilkinson *et al.*, 2017.
34 Bellinger & Sigee, 2015.
35 Bellinger & Sigee, 2015.
36 Money & Fischer, 2021.
37 Bellinger & Sigee, 2015.
38 Lewin, 1953.
39 Taylor, 1962, p. 584.
40 Taylor, 1962.
41 O'Hagan, 2002.
42 Vickerman & Sleigh, 2000.
43 Leidy, 1878, p. 235.
44 Warren, 1998.
45 Thomson, 2008, p. 6.
46 Leidy, 1879.
47 Mackinnon & Hawes, 1961.
48 Taylor, 1962.
49 Mackinnon & Hawes, 1961.
50 Mackinnon & Hawes, 1961.
51 Keeton, 1980.
52 Corliss, 1974.
53 Esteban & Fenchel, 2020.
54 Vickerman, 1992, p. 335.
55 Margulis & Chapman, 2009; Esteban & Fenchel, 2020.
56 Vandermeer, 1969.
57 Gause, 1934; Vandermeer, 1969.
58 Wilkinson, 2021.
59 Kang *et al.*, 2017; Lara *et al.*, 2020.
60 Lara *et al.*, 2020.
61 Chardez, 1986; Buckman *et al.*, 2024.
62 Hansell, 2007.
63 Eckert & McGee-Russell, 1974.
64 Armynot du Châtelet *et al.*, 2013.
65 Money & Fischer, 2021.
66 Dumack *et al.*, 2024.
67 Wilkinson & Mitchell, 2010.
68 Han *et al.*, 2008.
69 Wilkinson & Mitchell, 2010.
70 Yeates & Foissner, 1995.
71 Gomaa *et al.*, 2014.
72 Montagnes *et al.*, 2012; Van Etten *et al.*, 2023.

CHAPTER 5

1 Penard, 1902.
2 Corliss, 2001.
3 In the original French: '… après deux années d'un travail méthodique et poursuivi sans arrêt, j'ai put retrouver dans le bassin du Léman le 92% environ des espèces décrites dans le monde' (Penard, 1902, p. 10).
4 Tansley, 1939.
5 Finlay & Maberly, 2000.
6 Finlay & Maberly, 2000. The Institute of Freshwater Ecology later became part of the UK Centre for Ecology and Hydrology and moved away from its Windermere location.
7 Finlay & Maberly, 2000.
8 Esteban *et al.*, 2006.
9 Clarke, 2003.
10 Fenchel *et al.*, 1997.
11 O'Malley, 2014.
12 Fenchel *et al.*, 1997.
13 Baas Becking, 1934.
14 For the English translation, see Baas Becking, 2016.
15 De Wit & Bouvier, 2006.
16 O'Malley, 2007.
17 Sandon, 1927, p. 68.
18 Finlay, 2002.
19 Foissner, 2011.
20 Albertano, 2012.
21 Wilkinson, 2001.
22 Yang *et al.*, 2010.
23 Souffreau *et al.*, 2013.
24 Lahr *et al.*, 2014.
25 Harris & Yalden, 2008.
26 Wilkinson *et al.*, 2012.
27 Šanti-Temkiv *et al.*, 2022.
28 Rodó *et al.*, 2024.
29 Heger *et al.*, 2011a.
30 Kosakyan *et al.*, 2025.
31 Wilkinson, 2022.
32 Heger *et al.*, 2011a.
33 For any reader interested in the maths, we used a non-metric multidimensional scaling ordination.
34 Meisterfeld, 2008.
35 Hines *et al.*, 2020.
36 Hamilton & Lenton, 1998.
37 For background on Kathleen Atkinson I am indebted to her former colleague Liz Haworth, and her niece Hilary Atkinson.
38 Proctor, 1959.

39 Atkinson, 1972.
40 Green *et al.*, 2023.
41 Cochak *et al.*, 2026.
42 Manning *et al.*, 2021.
43 Sha *et al.*, 2025.
44 Kaštovský, 2024.
45 Bass & Boenick, 2011.
46 Bamforth, 1981.
47 For example, see Esteban & Finlay, 2008; Esteban & Fenchel, 2020.
48 Pinseel *et al.*, 2024.

CHAPTER 6

1 Matless, 1998.
2 Proctor *et al.*, 1996.
3 Bell, 1988.
4 From Act 4, Scene 6 in the RSC *Complete Works* (2007) which is based on the 1623 First Folio. There are two different surviving seventeenth-century versions of the play, so the quotation is sometimes to be found in Scene 7.
5 Farjon, 2023, p. 103.
6 Rackham, 2006, p. 44.
7 Schoenherr, 1992.
8 Lanner & Connor, 2001.
9 Garcia *et al.*, 2011.
10 Qiu *et al.*, 2021.
11 Galipaud & Kokko, 2020.
12 Silvertown & Lovett Doust, 1993.
13 Chepurnov *et al.*, 2004.
14 Galipaud & Kokko, 2020.
15 Speijer *et al.*, 2015.
16 Hamilton's papers on the evolution of sex are brought together in Hamilton (2001). See Wilkinson (2000) for an introduction to the history of the parasite-based theory of sex. Sherratt & Wilkinson (2009, Chapter 2) provide an introduction to the evolution of sexual reproduction, citing many of the key papers.
17 Judson & Normark, 1996; Speijer *et al.*, 2015.
18 Lahr *et al.*, 2011.
19 Speijer *et al.*, 2015.
20 Gladyshev *et al.*, 2008.
21 Wilson & Sherman, 2010.
22 Lacey, 2005, p. 190.
23 Flemming, 2008, p. 396.
24 Brandt *et al.*, 2015.
25 Lücking & Spribille, 2024.
26 Samba-Louaka, 2023.
27 For example, see Lipman, 1928.
28 Kokko, 2017.

CHAPTER 7

1 Roger Mason went on to have a long career as a geologist. Tina (C. L.) Negus was discouraged from following a geological career by her teachers, and instead studied combined zoology, botany and geography at university (before doing research on freshwater mussels), switching to art and poetry in later life. My brief summary of the history of the discovery is based on multiple sources, especially Brasier (2009).
2 Ford, 1958.
3 Attenborough, 2011 p. 34.
4 Gould, 1985, p. 232.
5 Kenchington *et al.*, 2018.
6 Bonner, 2019.
7 Witches' butter is still sometimes used as a common name for some fungi such as *Tremella mesenterica* and *Ezidia glandulosa*. This provides a good example of how common names can sometimes be very confusing, and hard to interpret.
8 Johnson, 1928.
9 Wolpert & Szathmáry, 2002.
10 Nave, 1867.
11 Kirk, 2004.
12 Canter-Lund & Lund, 1995.
13 Bonner, 1993.
14 Boraas *et al.*, 1998.
15 Maynard Smith & Szathmáry, 1995.
16 Umen & Herron, 2021, p. 624.
17 Romeralo *et al.*, 2012.

18 Stephenson & Feest, 2012.
19 Zhang *et al.*, 2024.
20 Bonner, 2009.
21 Bonner, 2009.
22 Swanson *et al.*, 1999.
23 Cavender, 2013.
24 Romeralo *et al.*, 2011.
25 Perrigo *et al.*, 2013.
26 For example, see Zhang *et al.*, 2024.
27 Strassmann *et al.*, 2000.
28 Tarnita *et al.*, 2015.
29 Brock *et al.*, 2011.
30 Requejo & Andrés-Rodríguez, 2019.
31 Rao & Chen, 2023.
32 Feest, 1996; Stephenson & Feest, 2012.
33 Ing, 2003.
34 Ing, 2025a.
35 Ing, 2011.
36 Ing, 2025b.
37 Wolpert & Szathmáry, 2002.
38 Jahan *et al.*, 2025.

CHAPTER 8

1 Information from Rackham's field notebook ('Red Notebook' number 242); available via the Cambridge Digital Library (https://cudl.lib.cam.ac.uk/collections/rackham).
2 Rackham, 2006.
3 Mabey, 1986; Dadswell, 2003.
4 In Book 3 of his *De causis plantarum*, one of the founding documents of botany (I have used the 1990 Loeb Classical Library Translation).
5 Thorne & Blandford, 2017.
6 Williams & Ducklow, 2019.
7 Azam *et al.*, 1983.
8 Fenchel, 2008, p. 192.
9 Gould, 2002, p. 875.
10 Bonkowski & Clarholm, 2012.
11 Geisen *et al.*, 2017.
12 Esteban & Fenchel, 2020.
13 Wilkinson, 2021, p. 125.
14 Lovelock, 1993.
15 van der Putten *et al.*, 2013, p. 272.
16 Cameron & Earley, 2015.
17 Wilkinson, 2025.
18 Geisen *et al.*, 2014.
19 Bardgett, 2005.
20 Maillard *et al.*, 2025.
21 Cohen *et al.*, 2003.
22 Yeates & Foissner, 1995.
23 Geisen *et al.*, 2015.
24 Schaller, 1974.
25 Small, 2024.
26 Geisen *et al.*, 2015.
27 A much more detailed and fully referenced account of the role of testate amoebae in nutrient cycling in soils, describing the literature up to 2009, is given by Wilkinson & Mitchell (2010).
28 Lousier & Parkinson, 1981.
29 Wilkinson & Mitchell, 2010.
30 Puppe, 2020.
31 For an example from a British nature reserve, see Creevy *et al.*, 2016.
32 Warming, 1909, p. 87.
33 Jassey *et al.*, 2022.
34 Lund, 1945.
35 Schmidt *et al.*, 2016.
36 Pentecost & Haworth, 2021.
37 Maberly *et al.*, 2022.
38 Maberly *et al.*, 2018.
39 Lund, 1950.
40 Pentecost, 2009.
41 The classic summary of these experiments is Schindler (1977), a research paper that has now been cited 3,790 times (Google Scholar data as of 31 December 2025).
42 Whitton & Potts, 2012.
43 Scheffer *et al.*, 2015.
44 Hutchinson, 1964, p. 336.
45 Doolittle, 2024.
46 Sogin *et al.*, 2006.
47 Liu *et al.*, 2015.
48 Pascoal *et al.*, 2021.
49 Weisse & Montagnes, 2022.
50 For example, see Margulis & Lovelock, 1974.

CHAPTER 9

1 White, 1789.
2 Latour, 2017, p. 8.
3 Latour, 2017, p. 7.
4 Hambler *et al.*, 2011.
5 Lamb, 1959.
6 Aboal *et al.*, 2016.
7 Mann & Droop, 1996.
8 Redford, 2023, p. 1.
9 Wilkinson, 2007.
10 'Free-living' is an important caveat here. Microbes such as parasites, that live in or on another species (technically known as its host), are not considered in this book, but if their host goes extinct they will follow unless they are able to utilise additional species as hosts. So every time you heard of a larger species becoming extinct, it's likely that several microbial species were lost at the same time (Wilkinson, 2007).
11 Cairns, 1993.
12 Stace & Crawley, 2015.
13 For example, see Heger *et al.*, 2011b.
14 Cox *et al.*, 2020, p. 57.
15 Smith & Wilkinson, 2007.
16 Bruni *et al.*, 2024.
17 Smith & Wilkinson, 2007.
18 Bruni *et al.*, 2024.
19 Cox *et al.*, 2020.
20 Wilkinson, 2010.
21 Perrigo *et al.*, 2012.
22 Mazei *et al.*, 2018.
23 Coulson *et al.*, 2013.
24 Mazei *et al.*, 2018, p. 14.
25 Wilkinson, 1994.
26 Amesbury *et al.*, 2017.
27 Gardner, 2018.
28 Li *et al.*, 2025.
29 Thomas, 2017.
30 Li *et al.*, 2025.
31 Mander *et al.*, 2025.
32 Rydin & Jeglum, 2013.
33 Lindsay, 1993.
34 Lawton, 1996, p. 209.
35 Evans *et al.*, 2024.
36 Creevy *et al.*, 2023; Liu *et al.*, 2025.
37 Jassey, 2025.
38 Hamard *et al.*, 2025.
39 Hamard *et al.*, 2021.
40 For example, see Creevy *et al.*, 2023.
41 Kokfelt *et al.*, 2009.
42 Kelly *et al.*, 2018.
43 For example, see Patrick, 1949.
44 Bellinger & Sigee, 2015.
45 Nisbet, 1991.
46 My discussion of diatoms and lake acidification draws heavily on an autobiographical essay by Battarbee (2015), along with my own memory of the debates at the time. Rick Battarbee and Roger Flower were key scientists involved in this work during the 1980s. Others included Kathleen Atkinson, whose work on waterbirds and algal dispersal was discussed in Chapter 5. Along with Liz Haworth, she compared the diatom record in Devoke Water (Lake District, influenced by atmospheric pollution) with Loch Sionascaig (Scotland, north of Ullapool, with much cleaner air). Acidity had increased in Devoke Water during the twentieth century but had not changed at Loch Sionascaig (Atkinson & Haworth, 1990).
47 Donoghue *et al.*, 2025.
48 Curtis, 2006.
49 O'Malley, 2014, p. 205.
50 Geisen *et al.*, 2017.

References

Aboal, M., Werner, O., García-Fernández, M. E., Palazazón, J. A., Cristóbal, J. C. & Williams, W. (2016) Should ecomorphs be conserved? The case of *Nostoc flagelliforme*, an endangered extremophile cyanobacteria. *Journal for Nature Conservation* 30, 52–64.

Albertano, P. (2012) Cyanobacterial biofilms in monuments and caves. In Whitton, B. A. (ed.), *Ecology of Cyanobacteria II*. Springer, Dordrecht, pp. 317–343.

Allen, D. E. (2010) *Books and Naturalists*. New Naturalist 112. Collins, London.

Allen, T. F. H. (1971) Multivariate approaches to the ecology of algae on terrestrial rock surfaces in North Wales. *Journal of Ecology* 59, 803–826.

Amesbury, M. J., Roland, T. P., Royles, J., Hodgson, D. A., Convey, P., Griffiths, H. & Charman, D. J. (2017) Widespread biological response to rapid warming on the Antarctic Peninsula. *Current Biology* 27, 1616–1622.

Andrews, J. H. (2017) *Comparative Ecology of Microorganisms and Macroorganisms*, 2nd edition. Springer, New York.

Antcliffe, J. B., Liu, A. G., Menon, L. R., McIlroy, D., McLoughlin, N. & Wacey, D. (2017) Understanding ancient life: how Martin Brasier changed the way we think about the fossil record. In Brasier, A. T., McIlroy, D. & McLoughlin, N. (eds), *Earth System Evolution and Early Life: a Celebration of the Work of Martin Brasier*. Geological Society of London, Special Publication 448, pp. 19–31.

Aristotle (1941) *The Basic Works of Aristotle* (ed. Richard McKeon). Random House, New York.

Armynot du Châtelet, E., Noiriel, C. & Delaine, M. (2013) Three-dimensional morphological and mineralogical characterization of testate amebae. *Microscopy and Microanalysis* 19, 1511–1522.

Atkinson, K. M. (1972) Birds as transporters of algae. *British Phycological Journal* 7, 319–321.

Atkinson, K. M. (1980) Experiments in dispersal of phytoplankton by ducks. *British Phycological Journal* 15, 49–58.

Atkinson, K. M. & Haworth, E. Y. (1990) Devoke Water and Loch Sionascaig: recent environmental changes and the post-glacial overview. *Philosophical Transactions of the Royal Society B* 327, 349–355.

Attenborough, D. (2011) *New Life Stories*. Collins, London.

Azam, F., Fenchel, T., Field, J. G., Gray, J. S., Meyer-Reil, L. & Thingstad, F. (1983) The ecological role of water-column microbes in the sea. *Marine Ecology Progress Series* 10, 257–263.

Baas Becking, L. G. M. (1934) *Geobiologie of inleiding tot de milieukunde*. W. P. Van Stockum & Zoon, The Hague.

Baas Becking, L. G. M. (2016) *Baas Becking's Geobiology*. Wiley Blackwell, Oxford. [Annotated English translation of Baas Becking, 1934.]

Bamforth, S. S. (1981) Protist biogeography. *Journal of Protozoology* 28, 2–9.

Bamforth, S. S. (2004) Water film fauna of microbiotic crusts of a warm desert. *Journal of Arid Environments* 56, 413–423.

Bardgett, R. (2005) The *Biology of Soil.* Oxford University Press, Oxford.

Barreno, E., Muggia, L., Chiva, S., Molins, A., Bordenave, C., Garcia-Breijo, F. & Moya, P. (2022) *Trebouxia* lynnae sp. nov. (former *Trebouxia* sp. TR9): biology and biogeography of an epitome lichen symbiotic microalga. *Biology* 11, 1196.

Bass, D. & Boenick, J. (2011) Everything is everywhere: a twenty-first century de-/reconstruction with respect to protists. In Fontaneto, D. (ed.), *Biogeography of Microscopic Organisms: Is Everything Small Everywhere?* Cambridge University Press, Cambridge, pp. 88–110.

Battarbee, R. W. (2015) Forestry, 'acid rain', and the acidification of lakes. In Thompson, D., Birks, H. & Birks, J. (eds), *Nature's Conscience: the Life and Legacy of Derek Ratcliffe.* Langford Press, Kings Lynn, pp. 385–400.

Beerling, D. (2019) *Making Eden.* Oxford University Press, Oxford.

Bell, G. (1988) *Sex and Death in Protozoa.* Cambridge University Press, Cambridge.

Bellinger, E. G. & Sigee, D. C. (2015) *Freshwater Algae: Identification, Enumeration and Use as Bioindicators,* 2nd edition. Wiley Blackwell, Chichester.

Belnap, J. (2003) The world at your feet: desert biological soil crusts. *Frontiers in Ecology and Environment* 1, 181–189.

Bhattacharya, D., Helmchen, T. & Melkonian, M. (1995) Molecular evolutionary analyses of nuclear-encoded small subunit ribosomal RNA identify an independent rhizopod lineage containing the Euglyphina and the Chlorarachniophyta. *Journal of Eukaryotic Microbiology* 42, 65–69.

Birch, T. (1756). *A History of the Royal Society of London for Improving Natural Knowledge, from its First Rise.* Vol. 2. A. Millar, London.

Boenigk, J., Wodniok, S. & Glucksman, E. (2015) *Biodiversity and Earth history.* Springer, Heidelberg.

Bolton, J. J. (2016) What is aquatic botany? And why algae are plants: the importance on non-taxonomic terms for groups of organisms. *Aquatic Botany* 132, 1–4.

Bonkowski, M. & Clarholm, M. (2012) Stimulation of plant growth through interactions of bacteria and protozoa: testing the auxiliary microbial loop hypothesis. *Acta Protozoologica* 51, 237–247.

Bonner, J. T. (1969) Hormones in social amoebae and mammals. *Scientific American* 220, 78–91.

Bonner, J. T. (1993) *Life Cycles.* Princeton University Press, Princeton.

Bonner, J. T. (2006) *Why Size Matters.* Princeton University Press, Princeton.

Bonner, J. T. (2009) *The Social Amoebae: the Biology of Cellular Slime Molds.* Princeton University Press, Princeton.

Bonner, J. T. (2013) *Randomness in Evolution.* Princeton University Press, Princeton.

Bonner, J. T. (2019) The evolution of evolution. *Journal of Experimental Zoology (Molecular and Developmental Evolution)* 332, 301–306.

Boraas, M. E., Seale, D. B. & Boxhorn, J. E. (1998) Phagotrophy by a flagellate selects for colonial prey: a possible origin of multicellularity. *Evolutionary Ecology* 12, 153–164.

Bower, F. O. (1935) *Primitive Land Plants.* Macmillan, London.

Brandt, A., de Vera, J.-P., Onofri, S. & Ott, S. (2015) Viability of the lichen *Xanthoria elegans* and its symbionts after 18 months of space exposure and simulated Mars conditions on the ISS. *International Journal of Astrobiology* 14, 411–425.

Brasier, A. T., McIlroy, D. & McLoughlin, N. (2017a) Contributions of Professor Brasier to the study of early life, stratigraphy and biochemistry. In Brasier, A. T., McIlroy, D. & McLoughlin, N. (eds), *Earth System Evolution and Early Life: a Celebration of the Work of Martin Brasier.* Geological Society of London, Special Publication 448, pp. 2–17.

Brasier, A. T., Culwick, T., Battison, L., Callow, R. H. T. & Brasier, M. D. (2017b) Evaluating evidence from the Torridonian Supergroup (Scotland, UK) for eukaryotic life on land in the Proterozoic. In Brasier, A. T., McIlroy, D. & McLoughlin, N. (eds),

Earth System Evolution and Early Life: a Celebration of the Work of Martin Brasier. Geological Society of London, Special Publication 448, pp. 121–144.

Brasier, M. (2009) *Darwin's Lost World.* Oxford University Press, Oxford.

Brasier, M. D. and 7 others (2002) Questioning the evidence for Earth's oldest fossils. *Nature* 416, 76–81.

Brock, D. A., Douglas, T. E., Queller, D. C. & Strassmann, J. E. (2011) Primitive agriculture in a social amoeba. *Nature* 469, 393–396.

Bronowski, J. (1968) Honest Jim and the tinker toy model. *The Nation*, 18 March, 381–382. [Reprinted in Stent, 1981.]

Brown, R. (1819) List of plants collected on the coasts of Baffin's Bay. In Ross, J., *A Voyage of Discovery ... for the Purpose of Exploring Baffin's Bay.* Longman, London, pp. cxli–cxliv.

Brown, R. (1866) *The Miscellaneous Botanical Works of Robert Brown.* Vol 1. The Ray Society, London.

Bruni, E. P. and 27 others (2024) Global distribution modelling of a conspicuous Gondwanian soil protist reveals latitudinal dispersal limitation and range contraction in response to climate warming. *Diversity and Distributions* 30, e13779.

Bryłka, K., Richoz, S., Alverson, A. J. & Conley, D. J. (2024) Looking for the oldest diatoms. *Marine Micropaleontology* 190, 102371.

Buckman, J., Krivtsov, V. & Poulton, A. J. (2024) An investigation into the morphological variation and ecological-environmental range of *Cyphoderia compressa*: a case study of Scottish material. *European Journal of Protistology* 95, 126106.

Burki, F., Roger, A. J., Brown, M. W. & Simpson, A. G. B. (2020) The new tree of eukaryotes. *Trends in Ecology and Evolution* 35, 43–55.

Cairns, J. (1993) Can microbial species with a cosmopolitan distribution become extinct? *Speculations in Science and Technology* 16, 69–73.

Cameron, L. & Earley, S. (2015) The ecosystem: movements, connections, tensions and translations. *Geoforum* 65, 473–481.

Cameron, I., Fyeff, B. & Kish, A. (2024) No Scottish snow patches survive until winter 2023/2024. *Weather* 79, 129–131.

Canter-Lund, H. & Lund J. (1995) *Freshwater Algae: Their Microscopic World Explored.* Biopress, Bristol.

Cavalier-Smith, T. (2010) Kingdoms Protozoa and Chromista and the eozoan root of the eukaryotic tree. *Biology Letters* 6, 342–345.

Cavender, J. C. (2013) A global overview of dictyostelid ecology with special emphasis on American forests. In Romeralo, M., Baldauf, S. & Escolante, R. (eds), *Dictyostelids.* Springer, Berlin, pp. 149–166.

Chardez, D. (1970) Etudes sur *Centropyxis aculeata* (Eherenburh) Stein (Protozoa Rhizopoda Testacea). *Bulletin des Recherches agronomiques de Gembloux* N. S. 5, 76–86.

Chardez, D. (1986) Thécamoebiens des plages de la Mer du Nord en Angleterre. *Acta Protozoologica* 25, 375–378.

Chepurnov, V. A., Mann, D. G., Sabbe, K. & Vyverman, W. (2004) Experimental studies on sexual reproduction in diatoms. *International Review of Cytology* 237, 91–154.

Christenhusz, M. J. M., Fay, M. F. & Chase, M. W. (2018) *Plants of the World.* Kew Publishing, Kew.

Clark, D. R. and 8 others (2018) Streams of data from drops of water: 21st century molecular ecology. *WIREs Water* 5, e1280.

Clarke, A. (2017) *Principles of Thermal Ecology.* Oxford University Press, Oxford.

Clarke, K. J. (2003) *Guide to the Identification of Soil Protozoa: Testate Amoebae.* Freshwater Biological Association, Ambleside.

Cobb, M. (2025) *Crick: a Mind in Motion – from DNA to the Brain.* Profile Books, London.

Cochak, C., Green, A. J., Sánchez, M. I., Pineda, A., de Oliveira de Matos, M. H. & Velho, L. F. M. (2026) Experimental evidence of ciliate dispersal and successional patterns via endozoochory in temporary aquatic systems. *Protist* 180, 126136.

Cohen, J. E., Jonsson, T. & Carpenter, S. R. (2003) Ecological community description using food web, species abundance, and body size. *Proceedings of the National*

Academy of Sciences of the USA 100, 1781–1786.

Corbet, S. A. (1973) An illustrated introduction to the testate rhizopods in *Sphagnum*, with special reference to the area around Malham Tarn, Yorkshire. *Field Studies* 3, 801–838.

Corliss, J. O. (1974) Classification and phylogeny of the Protista. *Actualités Protozoologiques* 1, 251–264.

Corliss, J. O. (2001) The two most remarkable *Amoeba* men: Joseph Leidy (1823–1891) of Philadelphia and Eugène Penard (1855–1954) of Geneva. *Protist* 152, 69–85.

Corliss, J. O. (2002) A salute to Antony van Leeuwenhoek of Delft, most versatile 17th century founding father of protistology. *Protist* 153, 177–90.

Coulson, S. J. and 12 others (2013) Introduction of invertebrates into the High Arctic via imported soils: the case of Barentsburg in Svalbard. *Biological Invasions* 15, 1–5.

Cox, C. B., Ladle, R. J. & Moore, P. D. (2020) *Biogeography: an Ecological and Evolutionary Approach*, 10th edition. Wiley Blackwell, Chichester.

Creevy, A. L. (2013) Microbial diversity of a nature reserve: eukaryotic microorganisms of Mere Sands Wood in N. W. England. MPhil thesis, Liverpool John Moores University.

Creevy, A., Fisher, J., Puppe, D. & Wilkinson, D. M. (2016) Protist diversity on a nature reserve in NW England – with particular reference to their role in soil biogenic silicon pools. *Pedobiologia* 59, 51–59.

Creevy, A. L., Wilkinson, D. M., Andersen, R. & Payne, R. J. (2023) Testate amoebae response and vegetation composition after plantation removal on a former raised bog. *European Journal of Protistology* 89, 125977.

Cunliffe, B. (2015) *By Steppe, Desert, and Ocean: the Birth of Eurasia*. Oxford University Press, Oxford.

Curtis, T. P. (2006) Microbial ecologists: it's time to 'go large'. *Nature Reviews Microbiology* 4, 488.

Dadswell, T. (2003) *The Selborne Pioneer: Gilbert White as Naturalist and Scientist: a Re-Examination*. Ashgate, Aldershot.

Dańko, M. J., Kozłowski, J. & Schaible, R. (2015) Unravelling the non-senescence phenomenon in *Hydra*. *Journal of Theoretical Biology* 382, 137–149.

Darwin, C. (1839) *Journal of Researches into the Geology and Natural History of the Various Countries Visited by H. M. S. Beagle Under the Command of Captain Fitzroy R. N. from 1832–1836*. Henry Colburn, London.

Davis, S. R. & Wilkinson, D. M. (2004) The conservation management value of testate amoebae as 'restoration' indicators: speculations based on two damaged raised mires in northwest England. *The Holocene* 14, 135–143.

Descamps, S. and 8 others (2017) Climate change impacts on wildlife in a High Arctic archipelago – Svalbard, Norway. *Global Change Biology* 23, 490–502.

de Vries, C., Galipaud, M. & Kokko, H. (2023) Extrinsic mortality and senescence: a guide for the perplexed. *Peer Community Journal* 3, e29.

de Wit, R. & Bouvier, T. (2006) 'Everything is everywhere, but, the environment selects': what did Baas Becking and Beijeinck really say? *Environmental Microbiology* 8, 755–758.

Donoghue, P. C. J., Lenton, T. M., Okasha, S., Shields, G. A. & Spang, A. (2025) Chance and purpose in the evolution of biospheres. *Philosophical Transactions of the Royal Society B* 380, 20240085.

Doolittle, W. F. (2024) *Darwinizing Gaia*. MIT Press, Cambridge, MA.

Dow, C. S. & Swoboda, U. K. (2000) Cyanotoxins. In Whitton, B. A. & Potts, M. (eds), *Ecology of Cyanobacteria*. Kluwer Academic Publishers, Dordrecht, pp. 613–632.

Duckert, C., Greeves, E., Blandemier, Q., Payne, R. & Mitchell, E. A. D. (2023) *Hyalosphenia papilio paynei* subsp. nov. – a highly conspicuous and localized *Sphagnum* peatland testate amoeba. *Acta Protozoologica* 62, 57–63.

Duigan, C. A., Allott, T. E. H., Monteith, D. T., Patrick, S. T., Lancaster, J. & Seda, J. M. (1998) The ecology and conservation of Llyn Idwal and Llyn Cwellyn (Snowdonia National Park, North Wales, UK) – two lakes proposed as special areas of conservation in Europe. *Aquatic Conservation: Marine and Freshwater Ecosystems.* 8, 325–336.

Dumack, K. and 8 others (2024) It's time to consider the Arcellinida shell as a weapon. *European Journal of Protistology* 92, 126051.

Dusenbery, D. B. (2009) *Living at Micro Scale.* Harvard University Press, Cambridge, MA.

Eckert, B. S. & McGee-Russell, S. M. (1974) Shell structure in *Difflugia lobostoma* observed by scanning and transmission electron microscopy. *Tissue and Cell* 6, 215–221.

Edwards, D. & Kenrick, P. (2015) The early evolution of land plants, from fossils to genomics: a commentary on Lang (1937) 'On the plant-remains from the Downtonian of England and Wales'. *Philosophical Transactions of the Royal Society B* 370, 20140343.

Edwards, D., Kenrick, P. & Dolen, L. (2017) History and contemporary significance of the Rhynie cherts: our earliest preserved terrestrial ecosystem. *Philosophical Transactions of the Royal Society B* 373, 20160489.

Eme, L. & Doolittle, W. F. (2015) Archaea. *Current Biology* 25, R851–R855.

Erktan, A., Or, D. & Scheu, S. (2020) The physical structure of soil: Determinant and consequence of trophic interactions. *Soil Biology and Biochemistry* 148, 107876.

Esteban, G. & Finlay, B. (2008) It's a small world. *British Wildlife* 19, 316–320.

Esteban, G. F. & Fenchel, T. M. (2020) *Ecology of Protozoa*, 2nd edition. Springer, Switzerland.

Esteban, G. F., Clarke, K. J., Olmo, J. L. & Finlay, B. J. (2006) Soil protozoa: an intensive study of population dynamics and community structure in an upland grassland. *Applied Soil Ecology* 33, 137–151.

Evans, C. R. C., Mullan, D. J., Roe, H. M., Fox, P. M., Gray, S. & Swindles, G. T. (2024) Responses of testate amoebae assemblages to peatland drain blocking. *Wetlands Ecology and Management* 32, 1–18.

Fara, P. (2009) A microscopic reality tale. *Nature* 459, 642–644.

Farjon, A. (2023) Oaks in J. G. Strutt's *Sylva Britannica* (1826) and their existence in 2013. *British Wildlife* 35, 95–103.

Feest, A. (1996) Slime moulds: stranger than fiction. *British Wildlife* 7, 236–242.

Fenchel, T. (2008) The microbial loop – 25 years later. *Journal of Experimental Marine Biology and Ecology* 366, 99–103.

Fenchel, T., Esteban, G. F. & Finlay, B. J. (1997) Local versus global diversity of microorganisms: cryptic diversity of ciliated protozoa. *Oikos* 80, 220–225.

Feynman, R. P., Leighton, R. B. & Sands, M. (1964) *The Feynman Lectures on Physics.* CALTECH, California.

Finlay, B. J. (2002) Global dispersal of free-living microbial eukaryote species. *Science* 296, 1061–1063.

Finlay, B. J. & Maberly, S. C. (2000) *Microbial Diversity in Priest Pot: a Productive Pond in the English Lake District.* Freshwater Biological Association, Ambleside.

Flemming, A. (2008) Bears in Britain. *British Wildlife* 19, 393–398.

Foissner, W. (2011) Dispersal of protists: the role of cysts and human introductions. In Fontaneto, D. (ed.), *Biogeography of Microscopic Organisms: Is Everything Small Everywhere?* Cambridge University Press, Cambridge, pp. 61–87.

Foissner, W., Agatha, S. & Berger, H. (2002) Soil ciliates (Protozoa, Ciliophora) from Namibia (Southwest Africa), with emphasis on two contrasting environments, the Etosha Region and the Namib Desert. *Denisia* 5, 1–1459.

Ford, B. J. (2007a) Antony van Leeuwenhoek: the discoverer of bacteria. In R. Huxley (ed.), *The Great Naturalists.* Thames and Hudson, London, pp. 104–110.

Ford, B. J. (2007b) Celebrating Leeuwenhoek's 375th birthday. What could his microscopes reveal? *Infocus* December 2007, 6–19.

Ford, T. D. (1958) Pre-Cambrian fossils from Charnwood Forest. *Proceedings of the Yorkshire Geological Society* 31, 211–217.

Friend, P. (2012) *Scotland.* New Naturalist 119. Collins, London.

Fry, N. and 7 others (2024) Living in the cracks: two novel genera of *Variosea*

(Amoebozoa) discovered on an urban sidewalk. *Journal of Eukaryotic Microbiology* 71, e13020.

Gabr, A., Grossman, A. R. & Bhattacharya, D. (2020) *Paulinella*, a model for understanding plastid primary endosymbiosis. *Journal of Phychology* 56, 837–843.

Galipaud, M. & Kokko, H. (2020) Sex, senescence, sources and sinks. *Functional Ecology* 34, 55–64.

Garcia, M. B., Dahlgen, J. P. & Ehrlén, J. (2011) No evidence for senescence in a 300-year-old mountain herb. *Journal of Ecology* 99, 1424–1430.

Gardner, D. K. (2018) *Environmental Pollution in China: What Everybody Needs to Know*. Oxford University Press, Oxford.

Gaston, K. J. (2023) *Common Ecology*. International Ecology Institute, Oldendorf/Luhe.

Gause, G. F. (1934) *The Struggle for Existence*. Williams & Wilkins, Baltimore.

Geisen, S., Bandow, C., Römbke, J. & Bonkowski, M. (2014) Soil water availability strongly alters the community composition of soil protists. *Pedobiologica* 57, 205–213.

Geisen, S., Rosengarten, J., Koller, R., Mulder, C., Urich, T. & Bonkowski, M. (2015) Pack hunting by a common soil amoeba on nematodes. *Environmental Microbiology* 17, 4538–4546.

Geisen S. and 46 others (2017) Soil protistology rebooted: 30 fundamental questions to start with. *Soil Biology and Biochemistry* 111, 94–103.

Gest, H. (2004) The discovery of microorganisms by Robert Hooke and Antoni van Leeuwenhoek, Fellows of the Royal Society. *Notes and Records of the Royal Society of London* 58, 187–201.

Gilbert, D., Amblard, C., Bourdier, G., Francez, A.-J. & Mitchell, E. A. D. (2000) Le régime alimentaire des Thécamoebiens (Protista, Sarcodina). *Année Biologique* 39, 57–68.

Gladyshev, E. A., Meselson, M. & Arkhipova, I. R. (2008) Massive horizontal gene transfer in bdelloid rotifers. *Science* 320, 1210–1213.

Golemansky, V. & Todorov, M. (2009) Testate amoebae (Arcellinida and Euglyphida) from the hydropsammon of Lake Leman (Switzerland). *Acta Zoologica Bulgarica* 61, 27–32.

Gomaa, F., Kosakyan, A., Heger, T. J., Corsaro, D., Mitchell, E. A. D. & Lara, E. (2014) One alga to rule them all: unrelated mixotrophic testate amoebae (Amoebozoa, Rhizaria and Stramenophiles) share the same symbiont (Trebouxiophyceae). *Protist* 165, 161–176.

González-Miguéns, R., Soler-Zamora, C., Villar-Depablo, M., Todorov, M. & Lara, E. (2022a) Multiple convergences in the evolutionary history of the testate amoebae family Arcellidae (Amoebozoa: Arcellinida: Sphaerothecinea): when the ecology rules the morphology. *Zoological Journal of the Linnean Society* 194, 1044–1071.

González-Miguéns, R. and 9 others (2022b) Deconstructing *Difflugia*: the tangled evolution of lobose testate amoebae shells (Amoebozoa: Arcellinida) illustrates the importance of convergent evolution in protist phylogeny. *Molecular Phylogenetics and Evolution* 175, 107557.

Gould, S. J. (1985) *The Flamingo's Smile*. W. W. Norton, New York.

Gould, S. J. (2002) *The Structure of Evolutionary Theory*. Harvard University Press, Cambridge, MA.

Green, A. J., Lovass-Kiss, Á., Reynolds, C., Sebastián-González, E., Silva, G. G., van Leeuwen, C. H. A. & Wilkinson, D. M. (2023) Dispersal of aquatic and terrestrial organisms by waterbirds: a review of current knowledge and future priorities. *Freshwater Biology* 68, 173–190.

Green, C. T. (1902) *The Flora of the Liverpool District*. D. Marples & Co., Liverpool.

Griffith, J. W. & Henfrey, A. (1875) *The Micrographic Dictionary*. John van Voorst, London.

Grube, M. & Balestrini, R. (2024) News from 'black belt' masters of symbiosis. *New Phytologist* 241, 1888–1890.

Guiry, M. D. (2024) How many species of algae are there? A reprise. Four kingdoms, 14 phyla, 63 classes and still growing. *Journal of Phycology* 60, 214–228.

Hagen, J. B. (2012). Five kingdoms, more or less: Robert Whittaker and the broad classification of organisms. *BioScience* 62, 67–74.

Haldane, J. B. S. (1927) *Possible Worlds and Other Essays*. Chatto & Windus, London.

Hamilton, W. D. (2001) *Narrow Roads of Geneland*, Volume 2. Oxford University Press, Oxford.

Hamilton, W. D. & Lenton, T. M. (1998) Spora and Gaia: how microbes fly with their clouds. *Ethology, Ecology and Evolution* 10, 1–16.

Hamard, S. and 12 others (2021) Contribution of microbial photosynthesis to peatland carbon uptake along a latitudinal gradient. *Journal of Ecology* 109, 3424–3441.

Hamard, S. and 18 others (2025) Microbial photosynthesis mitigates carbon loss from northern peatlands under warming. *Nature Climate Change* 15, 436–443.

Hambler, C., Henderson, P. A. & Speight, M. R. (2011) Extinction rates, extinction-prone habitats, and indicator groups in Britain and at larger scales. *Biological Conservation* 144, 713–721.

Han, B.-P., Wang, T., Lin, Q.-Q. & Dumont, H. J. (2008) Carnivory and active hunting by the planktonic testate amoebae *Difflugia tuberspinifera*. *Hydrobiologica* 596, 197–201.

Hansell, M. (2007) *Built By Animals: the Natural History of Animal Architecture*. Oxford University Press, Oxford.

Harris, S. & Yalden D. W. (eds) (2008) *Mammals of the British Isles: Handbook*, 4th edition. The Mammal Society, Southampton.

Heger, T. J. and 8 others (2011a) Rediscovery of *Nebela ansata* (Amoebozoa: Arcellinida) in eastern North America: biogeographical implications. *Journal of Biogeography* 38, 1897–1906.

Heger, T. J., Lara, E. & Mitchell, E. A. D. (2011b) Arcellida testate amoebae (Amoebozoa: Arcellinida): model of organisms for assessing microbial biogeography. In Fontaneto, D. (ed.), *Biogeography of Microscopic Organisms: Is Everything Small Everywhere?* Cambridge University Press, Cambridge, pp. 111–129.

Heger, T. J., Mitchell, E. A. D. & Leander, B. S. (2013) Holarctic phylogeography of the testate amoeba *Hylosphenia papilio* (Amoebozoa: Arcellinida) reveals extensive genetic diversity explained more by environment than dispersal. *Molecular Ecology* 22, 5172–5184.

Hines, H. N., McCarthy, P. J. & Esteban, G. F. (2020) First records of 'flagship' soil ciliates in North America. *Protist* 171, 125739.

Hingley, M. (1993) *Microscopic Life in Sphagnum*. Richmond Publishing, Slough.

Hooke, R. (1665) *Micrographia: or some physiological descriptions of minute bodies made by magnifying glasses with observations and inquiries thereupon*. The Royal Society, London.

Hutchinson, G. E. (1964) The lacustrine microcosm reconsidered. *American Scientist* 52, 334–341.

Ing, B. (2003) Snowline myxomycetes in Britain. *Field Mycology* 4, 7–14.

Ing, B. (2011) *Biodiversity in the North West: the Slime Moulds of Cheshire*. University of Chester Press, Chester.

Ing, B. (2020) *The Myxomycetes of Britain and Ireland*, Enlarged edition. Richmond Publishing, Slough.

Ing, B. (2025a) Snowline slime moulds – a group in danger. *British Wildlife* 37, 40–44.

Ing, B. (2025b) Slime moulds. *British Wildlife* 36, 302–303.

Jahan, I., Scott, T. J., Strassmann, J. E. & Queller, D. C. (2025) Testing the coordination hypothesis: incompatibilities in aggregative development of an experimentally evolved social amoeba. *Evolution Letters* 9, 236–248.

Jardine, L. (2003) *The Curious Life of Robert Hooke*. HarperCollins, London.

Jardine, L. (2008) *Going Dutch*. HarperCollins, London.

Jassey, V. (2025) Behind the paper. *Nature Climate Change* 15, 360–361.

Jassey, V. E. J. and 13 others (2015) An unexpected role for mixotrophs in the response of peatland carbon cycling to climatic warming. *Scientific Reports* 5, 16931.

Jassey, V. E. J. and 7 others (2022) Contribution of soil algae to the global carbon cycle. *New Phytologist* 234, 64–76.

John, D. M., Whitton, B. A. & Brook, A. J. (2011) *The Freshwater Algal Flora of the British Isles*, 2nd edition. Cambridge University Press, Cambridge.

John, D. M., Guiry, M. D., Wilbraham, J. & Krokowski, J. (2022) The 2011 edition of 'The freshwater algal flora of the British Isles': additions, corrections, nomenclatural and taxonomic changes. *Applied Phycology* 3, 36–71.

Johnson, W. (1928) *Gilbert White, Pioneer, Poet, and Stylist.* John Murray, London.

Jönsson, K. I., Rabbow, E., Schill, R. O., Harns-Ringdahl, M. & Rettberg, P. (2008) Tardigrades survive exposure to space in low Earth orbit. *Current Biology* 18, R729–R731.

Judson, O. P. & Normark, B. B. (1996) Ancient asexual scandals. *Trends in Ecology and Evolution* 11, 41–46.

Kang, S. and 14 others (2017) Between a pod and a hard place: the deep evolution of amoebae. *Molecular Biology and Evolution* 34, 2258–2270.

Kaštovský, J. (2024) Welcome to the jungle! An overview of modern taxonomy of cyanobacteria. *Hydrobiologica* 851, 1063–1077.

Katscher, F. (2004) The history of the terms prokaryotes and eukaryotes. *Protist* 155, 257–263.

Keeton, W. T. (1980) *Biological Science*, 3rd edition. W. W. Norton & Co., New York.

Kelly, M. (2000) *Identification of Common Benthic Diatoms in Rivers.* Field Studies Council AIDGAP Guides.

Kelly, M. G., Crisp, T. D., Lamb, B. & Whitton, B. (2018) Freshwater life. In Gater, S. (ed.), *The Natural History of Upper Teesdale*, 5th edition. Mosaic Teesdale Ltd, Middleton-in-Teesdale, pp. 165–178.

Kemp, M. (2006) *Seen/Unseen: Art, Science, and Intuition from Leonardo to the Hubble Telescope.* Oxford University Press, Oxford.

Kenchington, C. G., Harris, S. J., Vixseboxse, P. B., Pickup, C. & Wilby, P. R. (2018) The Ediacaran fossils of Charnwood Forest: shining new light on a major biological revolution. *Proceeding of the Geologists' Association* 129, 264–277.

Kington, J. (2010) *Climate and Weather.* New Naturalist 115. Collins, London.

Kirk, D. L. (2004) *Volvox. Current Biology* 14, R599–R600.

Kirkland, C. L. and 12 others (2025) A one-billion-year-old Scottish meteorite impact. *Geology* 53, 621–625.

Kokfelt, U., Struf, E. & Randsalu, L. (2009) Diatoms in peat – dominant producers in a changing environment? *Soil Biology and Biochemistry* 41, 1764–1766.

Kokko, H. (2017) Give one species the task to come up with a theory that spans them all: what good can come of that? *Proceedings of the Royal Society B* 284, 20171652.

Kosakyan, A., Gomaa, F., Lara, E. & Lahr, D. J. G. (2016) Current and future perspective on the systematics, taxonomy and nomenclature of testate amoebae. *European Journal of Protistology* 55, 105–117.

Kosakyan, A., Meisterfeld, R., Lara, E., Duckert, C. & Mitchell, E. A. D. (2025) *A Taxonomic Monograph of Hyalospheniid Testate Amoebae.* Alphil-Presses Universitaires Suisses, Neuchâtel.

Lacey, A. (2005) Death. In Honderich, T. (ed.), *The Oxford Companion to Philosophy*, 2nd edition. Oxford University Press, Oxford. pp 190–191.

Lahr, D. J. G., Parfrey, L. W., Mitchell, E. A. D., Katz, L. A. & Lara, E. (2011) The chastity of amoebae: re-evaluating evidence for sex in amoeboid organisms. *Proceedings of the Royal Society B* 278, 2081–2090.

Lahr, D. J., Lauginghouse IV, H. D., Oliverio, A. M., Gao, F. & Katz, L. A. (2014) How discordant morphological and molecular evolution among microorganisms can revise our notions of biodiversity on Earth. *Bioessays* 36, 950–959.

Lahr, D. J. G. and 20 others (2019) Phylogenomics and morphological reconstruction of Arcellinida testate amoebae highlight diversity of microbial eukaryotes in the Neoproterozoic. *Current Biology* 29, 991–1001.

Lamb, H. H. (1959) Our changing climate, past and present. *Weather* 14, 299–318.

Lane, N. (2015) The unseen world: reflections on Leeuwenhoek (1677) 'Concerning little animals'. *Philosophical Transactions of the Royal Society B* 370, 20140344.
Lane, N. (2017) Serial endosymbiosis or singular event at the origin of eukaryotes? *Journal of Theoretical Biology* 434, 58–67.
Lane, N. & Xavier, J. (2024) To unravel the origin of life, treat findings as pieces of a bigger puzzle. *Nature* 626, 948–951.
Lanner, R. M. & Connor, K. F. (2001) Does Bristlecone pine senesce? *Experimental Gerontology* 36, 675–685.
Lara, E., Dumavk, K., García-Martín, J. M., Kudryavtsev, A. & Kosakyan, A. (2020) Amoeboid protist systematics: A report on the 'Systematics of amoeboid protists' symposium at the VIIIth ECOP/ISOP meeting in Rome, 2019. *European Journal of Protistology* 76, 125727.
Latour, B. (2017) *Facing Gaia: Eight Lectures on the New Climatic Regime.* Translated by Catherine Porter. Polity Press, Cambridge.
Lawson, I. (2016) Crafting the microworld: how Robert Hooke constructed knowledge about small things. *Notes and Records of the Royal Society* 70, 23–44.
Lawton, J. H. (1996) Bog-plodging and sustainable development. *Oikos* 76, 209–210.
Lee, J. (2015) *Yorkshire Dales.* New Naturalist 130. Collins, London.
Leidy, J. (1878) *Amoeba proteus. American Naturalist* 12, 235–238.
Leidy, J. (1879) Fresh-water rhizopods of North America. *Report of the United States Geological Survey of the Territories.* 12, 1–324.
Lenton, T. M. & Daines, S. J. (2017) Matworld: the biogeochemical effects of early land life. *New Phytologist* 215, 531–537.
Lewin, J. C. (1953) Heterotrophy in diatoms. *Journal of General Microbiology* 9, 305–313.
Li, Z.-P. and 10 others (2025) Human land use promotes range expansion of soil protists from temperate to subtropical regions in China. *Proceedings of the National Academy of Sciences of the USA* 122, e2413220122.
Light, J. J. & Belcher, J. H. (1968) A snow microflora in the Cairngorm mountains of Scotland. *British Phycological Bulletin* 3, 471–473.
Lindsay, R. A. (1993) Peatland conservation: from cinders to Cinderella. *Biodiversity and Conservation* 2, 528–540.
Lipman, C. B. (1928) The discovery of living microorganisms in ancient rocks. *Science* 68, 272–231.
Liston, J. J. H. & Sanders, H. L. (2005) The 'other' Glasgow Boys: the rise and fall of a school of palaeobotany. In Bowden, A. J., Burek, C. V. & Wilding, R. (eds), *History of Palaeobotany: Selected Essays.* Geological Society of London, Special Publication 241, 197–228.
Liu, A. G., Menon, L. R., Shields, G. A., Callow, R. H. T. & McIlroy, D. (2017) Martin Brasier's contribution to the palaeobiology of the Ediacaran–Cambrian transition. In Brasier, A. T., McIlroy, D. & McLoughlin, N. (eds), *Earth System Evolution and Early Life: a Celebration of the Work of Martin Brasier.* Geological Society of London, Special Publication 448, pp. 179–193.
Liu, B., Heinemeyer, A., Marchant, R. & Mills, R. T. E. (2025) Exploring testate amoebae as taxonomic and functional bioindicators to inform peatland habitat status and blanket bog restoration. *Soil Biology and Biochemistry* 206, 109790.
Liu, L., Yang, J., Yu, Z. & Wilkinson, D. M. (2015) The biogeography of abundant and rare bacterioplankton in the lakes and reservoirs of China. *The ISME Journal* 9, 2068–2077.
Lousier, J. D. & Parkinson, D. (1981) The disappearance of empty tests of litter- and soil-testate amoebae (Testacea, Rhizopoda, Protozoa). *Archiv für Protistenkunde* 124, 312–336.
Lovelock, J. E. (1993) The soil as a model for the Earth. *Geoderma* 57, 213–215.
Lücking, R. & Spribille, T. (2024) *The Lives of Lichens.* Princeton University Press, Princeton.
Lund, J. W. G. (1945) Observations on soil algae. I. The ecology, size and taxonomy of British soil diatoms. *New Phytologist* 44, 196–219.
Lund, J. W. G. (1950) Studies on *Asterionella formosa* Hass. II. Nutrient depletion and the spring maximum. *Journal of Ecology* 38, 15–35.

Lund, J. W. G. (1961) The algae of the Malham Tarn district. *Field Studies* 1, 85–119.

Mabberley, D. J. (2017) *Mabberley's Plant-Book*, 4th edition. Cambridge University Press, Cambridge.

Maberly, S. C., Ciar, D., Elliott, J. A., Jones, I. D., Reynolds, C. S., Thackery, S. J. & Winfield, I. J. (2018) From ecological informatics to the generation of ecological knowledge: long-term research in the English Lake District. In Recknagel, F. & Michener, W. K. (eds), *Ecological Informatics*. Springer.

Maberly, S. C., Chao, A. & Finlay, B. J. (2022) Seasonal patterns of phytoplankton taxon richness in lakes: effects of temperature, turnover and abundance. *Protist* 173, 125925.

Mabey, R. (1986) *Gilbert White*. Collins, London.

MacGabhann, B. A. (2007) Discoidal fossils of the Ediacaran biota: a review of current understanding. In Vickers-Rich P. and Komarower P. (eds), *The Rise and Fall of the Ediacaran Biota*. Geological Society of London, Special Publication 286, 297–313.

Mackinnon, D. I. & Hawes, R. S. J. (1961) *An Introduction to the Study of Protozoa*. Clarendon Press, Oxford.

Macorano, L. & Nowack, E. C. M. (2021) *Paulinella chromatophora*. *Current Biology* 31, R1024–R1026.

Maillard, F. and 9 others (2025) Hidden decomposers: revisiting saprotrophy among soil protists and its potential impact on carbon cycling. *Soil Biology and Biochemistry* 205, 109786.

Mander, Ü., Öpik, M. & Espenberg, M. (2025) Global peatland greenhouse gas dynamics: state of the art, processes, and perspectives. *New Phytologist* 246, 94–102.

Mann, D. G. & Droop, S. J. M. (1996) Biodiversity, biogeography and conservation of diatoms. *Hydrobiologia* 336, 19–32.

Mann, D. G. & Vanormelingen, P. (2013) An inordinate fondness? The number, distributions, and origins of diatom species. *Journal of Eukaryotic Microbiology* 60, 414–420.

Manning, F. S., Curtis, P. J., Walker, I. R. & Pither, J. (2021) Potential long-distance dispersal of freshwater diatoms adhering to waterfowl plumage. *Freshwater Biology* 66, 1136–1148.

Margulis, L. (1971) The origin of plant and animal cells. *American Scientist* 59, 230–235.

Margulis, L. (1998) *The Symbiotic Planet*. Weidenfeld & Nicolson, London.

Margulis, L. & Chapman, M. J. (2009) *Kingdoms and Domains*. Academic Press, Amsterdam.

Margulis, L. & Lovelock, J. E. (1974) Biological modulation of the Earth's atmosphere. *Icarus* 21, 471–489.

Marin, B., Nowack, E. C. M. & Melkonian, M. (2005) A plastid in the making: evidence for a second primary endosymbiosis. *Protist* 156, 425–432.

Martin, W. F., Tielens, A. G. M., Mental, M., Garg, S. G. & Gould, S. B. (2017) The physiology of phagocytosis in the context of mitochondrial origin. *Microbiology and Molecular Biology Reviews* 81, e00008–17.

Matless, D. (1998) *Landscape and Englishness*. Reaktion Books, London.

Maynard Smith, J. (1968) *Mathematical Ideas in Biology*. Cambridge University Press, Cambridge.

Maynard Smith, J. & Szathmáry, E. (1995) *The Major Transitions in Evolution*. W. H. Freeman, Oxford.

Mayr, E. (1982) *The Growth of Biological Thought*. Harvard University Press, Cambridge, MA.

Mazei, Y. A., Lebedeva, N. V., Taskaeva, A. A., Ivanovsky, A. A., Chernyshov, V. A., Tsyganov, A. N. & Payne, R. J. (2018) What role does human activity play in microbial biogeography? The revealing case of testate amoebae in the soils of Pyramiden, Svalbard. *Pedobiologica – Journal of Soil Ecology* 67, 10–15.

McGovern, S. T., Evans, C. D., Dennis, P., Walmsley, C. A., Turner, A. & McDonald, M. A. (2014) Increased inorganic nitrogen leaching from mountain grassland ecosystem following grazing removal: a hangover of past intensive land use? *Biogeochemistry* 119, 125–138.

Meisterfeld, R. (2008) *Bullinularia foissneri* nov. sp. (Amoebozoa, Arcellinida) from Australia and synopsis of the genus *Bullinularia*. *Denisia* 32, 235–241.

Melkonian, M. & Mollenhauer, D. (2005) Robert Lauterborn (1869–1952) and his *Paulinella chromatophora*. *Protist* 156, 253–262.

Merton, R. K. (1968) The Matthew effect in science. *Science* 159, 56–63.

Miao, L., Yin, Z., Knoll, A. H., Qu, Y. & Zhu, M. (2024) 1.63-billion-year-old multicellular eukaryotes from the Chuanliggou Formation in North China. *Science Advances* 10, eadk3208.

Midgley, M. (1995) *Beast and Man*, revised edition. Routledge, London.

Mitchell, E. A. D., Arnet, T., Chatelain, A., Derungs, N. & Lara, E. (2013) Thécamoebiens muscicoles d'un mur d'enceinte du Château de Neuchâtel – les protistes constituent-ils une partie néglibable de la biodiversité ou plutôt une 'majorité silencieuse'. *Bulletin de la Société Neuchâteloise Sciences Naturelles* 133, 83–92.

Mitchell, R. L., Strullu-Derrien, C., Sykes, D., Pressel, S. Duckett, J. G. & Kenrick, P. (2021) Cryptogamic ground covers as an analogue for early terrestrial biospheres: initiation and evolution of biologically mediated proto-soils. *Geobiology* 19, 292–306.

Money, N. P. (2014) *The Amoeba in the Room*. Oxford University Press, Oxford.

Money, N. P. & Fischer, M. W. F. (2021) What is the weight of a single amoeba and why does it matter? *The American Biology Teacher* 83, 571–574.

Montagnes, D., Roberts, E., Lukeš, J. & Lowe, C. (2012) The rise of model protozoa. *Trends in Microbiology* 20, 184–191.

Monteith, J. L. & Unsworth, M. H. (2008) *Principles of Environmental Physics*, 3rd edition. Academic Press, Amsterdam.

Moss, B. (2017) *Ponds and Small Lakes: Microorganisms and Freshwater Ecology*. Pelagic Publishing, Exeter.

Murray, W. H. (1951) *Undiscovered Scotland*. J. M. Dent & Son, London.

Murray, W. H. (1987) *Scotland's Mountains*. Scottish Mountaineering Trust, Edinburgh.

Nave, J. (1867) *A Handy-Book to the Collection and Preparation of Freshwater and Marine Algae, Diatoms, Desmids, Fungi, Lichens and Mosses*. Translated and edited by W. W. Spicer. Robert Hardwicke, London.

Nicholson, E. M. (1957) *Britain's Nature Reserves*. Country Life, London.

Nielsen, U. N. (2019) *Soil Faunal Assemblages*. Cambridge University Press, Cambridge.

Nisbet, E. G. (1989) Some northern sources of atmospheric methane: production, history, and future implications. *Canadian Journal of Earth Sciences* 26, 1603–1611.

Nisbet, E. G. (1991) *Leaving Eden: to Protect and Manage the Earth*. Cambridge University Press, Cambridge.

Nomura, M. & Ishida, K. (2016) Fine-structural observations on siliceous scale production and shell assembly in testate amoebae *Paulinella chromatophora*. *Protist* 167, 303–318.

Nomura, M., Nakayama, T. & Ishida, K. (2014) Detailed processes of shell construction in the photosynthetic testate amoeba *Paulinella chromatophora* (Euglyphid, Rhizaria). *Journal of Eukaryotic Microbiology* 61, 317–321.

Nowack, E. C. M. (2014) *Paulinella chromatophora*: rethinking the transition from endosymbiont to organelle. *Acta Societatis Botanicorum Poloniae* 83, 387–397.

Ogden, C. G. & Hedley, R. H. (1980) *An Atlas of Freshwater Testate Amoebae*. British Museum (Natural History) and Oxford University Press, Oxford.

O'Hagan, F. (2002) The contribution of religious orders to education in Glasgow during the period 1847–1918. PhD thesis, University of Glasgow.

Olby, R. (2009) *Francis Crick: Hunter of Life's Secrets*. Cold Spring Harbor Press, New York.

O'Malley, M. A. (2007) The nineteenth century roots of 'everything is everywhere'. *Nature Reviews Microbiology* 5, 647–651.

O'Malley, M. A. (2014) *Philosophy of Microbiology*. Cambridge University Press, Cambridge.

Osborne, A. W., Marsden, S., Caporn, S. J. M. & Coulthard, E. (2024) Cutting peat: the historical ecology and dissection on the Chat Moss ecosystem. *Mires and Peat* 31, 21.

Pace, N., Sapp, J. & Goldenfeld, N. (2012) Phylogeny and beyond: Scientific, historical, and conceptual significance of the first tree of life. *Proceedings of the National Academy of Sciences of the USA* 109, 1011–1018.

Pascoal, F., Costa, R. & Magalhães, C. (2021) The microbial rare biosphere: current concepts, methods and ecological principles. *FEMS Microbial Ecology* 97, fiaa227.

Patrick, R. (1949) A proposed biological measure of stream conditions, based on a survey of the Conestoga River, Lancaster County, Pennsylvania. *Proceedings of the Academy of Natural Sciences of Philadelphia* 1012, 227–341.

Pearsall, W. H. & Pennington, W. (1973) *The Lake District: a Landscape History*. New Naturalist 53. Collins, London.

Penard, E. (1902) *Faune rhizopodique du bassin du Léman*. Henry Kündig, Geneva.

Pentecost, A. (2009) The marl lakes of the British Isles. *Freshwater Reviews* 2, 167–197.

Pentecost, A. (2010) Some 'lamp floras' from tourist caves in northern England. *Cave and Karst Science* 37, 93–98.

Pentecost, A. & Haworth, E. (2021) *Freshwater Algae of Cumbria*. Freshwater Biological Association Scientific Publication 72.

Pentecost, A. & Whitton, B. A. (2012) Subaerial cyanobacteria. In B. A. Whitton (ed.), *Ecology of Cyanobacteria II*. Springer, Dordrecht, pp. 291–316.

Perrigo, A. L., Romeralo, M. & Baldauf, S. L. (2012) What's on your boots: an investigation into the role we play in protist dispersal. *Journal of Biogeography* 39, 998–1003.

Perrigo, A. L., Baldauf, S. L. & Romeralo, M. (2013) Diversity of dictyostelid social amoebae in high latitude habitats of northern Sweden. *Fungal Diversity* 58, 185–198.

Phillips, J. (2020) *Exploring Woven Fabrics*. Natural Timeout Publications.

Pigott, M. E. & Pigott, C. D. (1959) Stratigraphy and pollen analysis of Malham Tarn and Tarn Moss. *Field Studies* 1, 84–101.

Pinseel, E., Sabbe, K., Verleyen, E. & Vyverman, W. (2024) A new dawn for protist biogeography. *Global Ecology and Biogeography* 33, e13925.

Porfirio-Sousa, A. L. and 13 others (2024) Amoebozoan testate amoebae illuminate the diversity of heterotrophs and the complexity of ecosystems throughout geological time. *Proceedings of the National Academy of Sciences of the USA* 121, e2319628121.

Porter, S. M. (2016) Tiny vampires in ancient seas: evidence for predation via perforation in fossils from the 730–740-million-year-old Chuar Group, Grand Canyon, USA. *Proceedings of the Royal Society B* 283, 20160221.

Porter, S. M. & Knoll, A. H. (2000) Testate amoebae in the Neoproterozic Era: evidence from vase-shaped microfossils in the Chuar Group, Grand Canyon. *Paleobiology* 26, 360–385.

Porter, S. M., Meisterfeld, R. & Knoll, A. H. (2003) Vase-shaped microfossils from the Neoproterozoic Chuar Group, Grand Canyon: A classification guided by modern testate amoebae. *Journal of Paleontology* 77, 409–429.

Priddle, J. & Happey-Wood, C. M. (1983) Significance of small species of Chlorophyta in freshwater phytoplankton communities with special reference to five Welsh lakes. *Journal of Ecology* 71, 793–810.

Procházková, L., Leya, T., Křížková, H. & Nedbalová, L. (2019) *Sanguina nivaloides* and *Sanguina aurantia* gen. et spp. nov. (Chlorophyta): the taxonomy, phylogeny, biogeography and ecology of two newly recognised algae causing red and orange snow. *FEMS Microbial Ecology* 95, fiz064.

Proctor, M., Yeo, P. & Lack, A. (1996) *The Natural History of Pollination*. New Naturalist 83. Collins, London.

Proctor, V. W. (1959) Dispersal of fresh-water algae by migratory birds. *Science* 130, 623–624.

Puppe, D. (2020) Review on protozoic silica and its role in silicon cycling. *Geoderma* 365, 114224.

Qiu, T. and 60 others (2021) Is there tree senescence? The fecundity evidence. *Proceedings of the National Academy of Sciences of the USA* 118, e2106130118.

Rackham, O. (2006) *Woodlands.* New Naturalist 100. Collins, London.

Ralfs, J. (1848) *The British Desmidieae.* Reeve, Benham & Reeve, London.

Rao, G. & Chen, S. (2023) Advancements in the taxonomic study of myxomycetes (Myxogastrea) in China. *Mycology* 14, 316–325.

Redford, K. H. (2023) Extending conservation to include Earth's microbiome. *Conservation Biology* 37, e14088.

Requejo, O. & Andrés-Rodríguez, N. F. (2019) Consideraciones etnobiológicas sobre los mixomicetos. *Boletin de la Sociedad Micrológica de Madrid* 43, 25–37.

Rodó, X. and 13 others (2024) Microbial richness and air chemistry in aerosols above the PBL confirm 2,000-km long-distance transport of potential human pathogens. *Proceedings of the National Academy of Sciences of the USA* 121, e2404191121.

Romeralo, M., Moya-Larano, J. & Lado, C. (2011) Social amoebae: environmental factors influencing their distribution and diversity across South-Western Europe. *Microbial Ecology* 61, 154–165.

Romeralo, M., Escalante, R. & Baldauf, S. L. (2012) Evolution and diversity of Dictyostelid social amoebae. *Protist* 163, 327–343.

Rothschild, L. J. (1989) Protozoa, Protista, protoctista: what's in a name? *Journal of the History of Biology* 22, 277–305.

Ruggiero, M. A. and 8 others (2015) A higher level classification of all living organisms. *PLOS One* 10(4), e0119248.

Ruxton, G. D. (2019) *Nature's Giants.* Yale University Press, New Haven.

Rydin, H. & Jeglum, J. K. (2013) *The Biology of Peatlands,* 2nd edition. Oxford University Press, Oxford.

Sagan, L. (1967) On the origin of mitosing cells. *Journal of Theoretical Biology* 14, 225–274.

Samba-Louaka, A. (2023) Encystment of free-living amoebae, so many blind spots to cover. *Parasitologia* 3, 53–58.

Sanders, W. B. (2021) The photoaerogens: algae and plants reunited conceptually. *American Journal of Botany* 109, 363–365.

Sanderson, T. A. & Wilkinson, D. M. (1989) Notes on coleoptera around the margins of four upland ponds in mid Wales. *Entomologists Gazette* 40, 39–41.

Sandon, H. (1927) *The Composition and Distribution of the Protozoan Fauna in Soil.* Oliver & Boyd, Edinburgh.

Šanti-Temkiv, T., Amato, P., Casamayor, E. O., Lee, P. K. H. & Pointing, S. B. (2022) Microbial ecology of the atmosphere. *FEMS Microbiology Reviews* 46, 1–18.

Sapp, J. (2009) *The New Foundations of Evolution.* Oxford University Press, Oxford.

Saville Kent, W. (1880–1882) *A Manual of the Infusoria.* Three volumes. David Bogue, London.

Schaller, G. B. (1974) *Golden Shadows, Flying Hooves.* Collins, London.

Scheffer, M. and 12 others (2015) Creating a safe operating space for iconic ecosystems. *Science* 347, 1317–1319.

Schindler, D. W. (1977) Evolution of phosphorus limitation in lakes. *Science* 195, 260–262.

Schmidt, O., Dyckmans, J. & Schrader, S. (2016) Photoautotrophic microorganisms as carbon source for temperate soil invertebrates. *Biology Letters* 12, 20150646.

Schoenherr, A. A. (1992) *A Natural History of California.* California University Press, Berkeley.

Segawa, T. and 7 others (2018) Bipolar dispersal of red-snow algae. *Nature Communications* 9, 3094.

Sha, Z. and 8 others (2025) Evidence of global dispersal of the harmful cyanobacterium *Raphidiopis raciborskii*, in lentic freshwaters through migratory waterbirds. *Harmful Algae* 142, 102786.

Sharples, R. W. & Minter, S. W. (1983) Theophrastus on fungi: inaccurate citations in Athenaeus. *Journal of Hellenic Studies* 103, 154–156.

Sherratt, T. N. & Wilkinson, D. M. (2009) *Big Questions in Ecology and Evolution.* Oxford University Press, Oxford.

Silva, P. C., Mattox, K. R. & Blackwell, W. H. (1972) The generic name *Hormidium* as applied to green algae. *Taxon* 21, 639–645.

Silvertown, J. (2013) *The Long and the Short of It.* University of Chicago Press, Chicago.

Silvertown J. & Lovett Doust, J. (1993) *Introduction to Plant Population Biology*, 3rd edition. Blackwell Science, Oxford.

Small, R. W. (2024) An introduction to free-living soil nematodes. *British Wildlife* 36, 185–193.

Smith, H. G. (1992) Distribution and ecology of the testate rhizopod fauna of the continental Antarctic zone. *Polar Biology* 12, 629–634.

Smith, H. G. & Wilkinson, D. M. (2007) Not all free-living microorgansims have cosmopolitan distributions: the case of *Nebela (Apodera) vas* Certes (Protozoa: Amoebozoa: Arcellinida). *Journal of Biogeography* 34, 1822–1831.

Sogin, M. L. and 7 others (2006) Microbial diversity in the deep sea and underexplored 'rare biosphere'. *Proceedings of the National Academy of Sciences of the USA* 103, 12115–12120.

Soler-Zamora, C., Useros, F., González-Miguéns, R., Gómez-Rodriguez, P. & Lara, E. (2023) The problem of 'shadow species' as illustrated with the taxonomic hotchpotch *Cyphoderia ampulla* (Rhizaria: Cyphoderiidae). *Zoological Journal of the Linnean Society* 199, 477–492.

Souffreau, C., Vanormelingen, P., Van der Vijver, B., Isheva, T., Verleyen, E., Sabbe, K. & Vyverman, W. (2013) Molecular evidence for distinct Antarctic lineages in the cosmopolitan terrestrial diatoms *Pinnularia borealis* and *Hantzschia amphioxys. Protist* 164, 101–115.

Speijer, D., Lukeš, J. & Eliáš, M. (2015) Sex is a ubiquitous, ancient, and inherent attribute of eukaryotic life. *Proceedings of the National Academy of Sciences of the USA* 112, 8827–8834.

Stace, C. (2019) *New Flora of the British Isles*, 4th edition. C&M Floristics, Middlewood Green.

Stace, C. A. & Crawley, M. J. (2015) *Alien Plants.* New Naturalist 129. William Collins, London.

Stancheva, R., Hall, J. D., Herburger, K., Lewis, L. A., McCourt, R. M., Sheath, R. G. & Holzinger, A. (2014) Phylogenetic position of *Zygogonium ericetorum* (Zygenmatophyceae, Charophyta) from a high alpine habitat and ultrastructural characterization of unusual aplanospores. *Journal of Phycology* 50, 790–803.

Stent, G. S. (ed.) (1981) *The Double Helix: a New Critical Edition of J. D. Watson's Famous DNA Story.* Weidenfeld & Nicolson, London.

Stephenson, S. L. & Feest, A. (2012) Ecology of soil eumycetozoans. *Acta Protozoologica* 51, 201–208.

Strassmann, J. E., Zhu, Y. & Queller, D. C. (2000) Altruism and social cheating in the social amoeba *Dictyostelium discoideum. Nature* 408, 965–967.

Stroh, P. A., Walker, K. J., Humphery, T. A., Pescott, O. L. & Burkmar, R. J. (2023) *Plant Atlas 2020.* Princeton University Press, Princeton.

Strullu-Derrien, C., Kenrick, P. & Knoll, A. H. (2019a) The Rhynie chert. *Current Biology* 29, R1211–R1223.

Strullu-Derrien, C., Kenrick, P., Goral, T. & Knoll, A. H. (2019b) Testate amoebae in the 407-million-year-old Rhynie chert. *Current Biology* 29, 461–467.

Strullu-Derrien, C. and 7 others (2026) An arbuscular mycorrhiza from the 407-million-year-old Windyfield Chert identified through advanced fluorescence Raman imaging. *New Phytologist* 249, 448–459.

Subramanian, S. (2020) *A Dominant Character: the Radical Science and Restless Politics of J. B. S. Haldane.* Atlantic Books, London.

Swanson, A. R., Vadell, E. M. & Cavendert, J. C. (1999) Global distribution of soil dictyostelids. *Journal of Biogeography* 26, 133–148.

Tansley, A. G. (1935) The use and abuse of vegetation concepts and terms. *Ecology* 16, 284–307.

Tansley, A. G. (1939) *The British Islands and Their Vegetation.* Cambridge University Press, Cambridge.

Tarnita, C. E., Washburne, A., Martinez Garcia, R., Sgro, A. E. & Levin, S. A. (2015) Fitness tradeoffs between spores and nonaggregating cells can explain the coexistence of diverse genotypes in cellular slime molds. *Proceedings of the National Academy of Sciences of the USA* 112, 2776–2781.

Taylor, M. (1962) Ways and waywardness of amoebae. *American Scientist* 50, 576–596.

Thomas, C. D. (2017) *Inheritors of the Earth.* Allen Lane, London.

Thomas, J. V. & Buckton, S. J. (2023) If you go down to the woods today, you're sure of a small surprise… a tardigrade survey in and around York. *The Naturalist* 148, 29–37.

Thompson, A. R., Powell, G. S. & Adams, B. J. (2019) Provisional checklist of terrestrial heterotrophic protists from Antarctica. *Antarctic Science* 31, 287–303.

Thomson, K. (2008) *The Legacy of the Mastodon.* Yale University Press, New Haven & London.

Thorne, K. S. & Blandford, R. D. (2017) *Modern Classical Physics.* Princeton University Press, Princeton.

Umen, J. & Herron, M. D. (2021) Green algal models for multicellularity. *Annual Review of Genetics* 55, 603–632.

Useros, F., Garcia-Cunchillos, I., Henery, N., Berney, C. & Lara, E. (2024) How good are global DNA-based environmental surveys for detecting all protest diversity? Arcellinida as an example of biased representation. *Environmental Microbiology* 26, e16606.

van Damme, K. & Kotov, A. A. (2016) The fossil record of the Cladocera (Crustacea: Branchipoda): evidence and hypotheses. *Earth-Science Reviews* 163, 162–189.

Vandermeer, J. H. (1969) The competitive structure of communities: an experimental approach with protozoa. *Ecology* 50, 362–371.

van der Putten and 12 others (2013) Plant–soil feedbacks: the past, the present and future challenges. *Journal of Ecology* 101, 265–276.

van Dijk, P. J. & Ellis, T. H. N. (2020) Mendel's journey to Paris and London: context and significance for the origin of genetics. *Folia Mendeliana* 56, 5–34.

van Etten, J., Benites, L. F., Stephens, T. G., Yoon, H. S. & Bhattacharya, D. (2023) Algae obscura: the potential of rare species as model systems. *Journal of Phycology* 59, 293–300.

van Leeuwenhoek, A. (1674) More observations from Mr Leewenhook, in a letter of 7 September 1674, sent to the publisher. *Philosophical Transactions of the Royal Society* 12, 821–831.

Vickerman, K. (1992) The diversity and ecological significance of protozoa. *Biodiversity and Conservation* 1, 334–341.

Vickerman, K. & Sleigh, M. (2000) *A Century of Protozoology in Britain.* The British section of the Society of Protozoologists.

Vincent, W. F. (1988) *Microbial Ecosystems of Antarctica.* Cambridge University Press, Cambridge.

Vogel, S. (2013) *Comparative Biomechanics,* 2nd edition. Princeton University Press, Princeton.

Wanner, M., Nähring, J. M. & Fischer, R. (1997) Molecular identification of clones of testate amoebae using single nuclei for PCR amplification. *European Journal of Protistology* 33, 192–199.

Warming, E. (1890) *Handbuch der systematischen Botanik.* Gebrüder Borntraeger, Berlin.

Warming, E. (1909) *Oecology of Plants: an Introduction to the Study of Plant-Communities.* Clarendon Press, Oxford.

Warren, L. (1998) *Joseph Leidy: the Last Man Who Knew Everything.* Yale University Press, New Haven & London.

Weisse, T. & Montagnes, D. J. S. (2022) Ecology of planktonic ciliates in a changing world: Concepts, methods, and challenges. *Journal of Eukaryotic Microbiology* 69, e12879.

West, G. S. (1911) The British freshwater phytoplankton. In Tansley, A. G. (ed.), *Types of British Vegetation.* Cambridge University Press, Cambridge, pp. 196–203.

Westall, F. (2025) What the earliest evidence for life tells us about the early evolution of the biosphere. *Philosophical Transactions of the Royal Society B* 380, 20240106.

Wier, A. & Margulis, L. (2000) The wonderful lives of Joseph Leidy (1823–1891). *International Microbiology* 3, 55–58.

White, G. (1789) *The Natural History and Antiquities of Selborne.* B. White and Son, London.

Whitfield, C. J., Baulch, H. M., Chun, K. P. & Westbrook, C. J. (2015) Beaver-mediated methane emissions: the effects of population growth in Eurasia and the Americas. *Ambio* 44, 7–15.

Whitton, B. A. & Potts, M. (2000) *The Ecology of Cyanobacteria: Their Diversity in Time and Space.* Kluwer Academic Publishers, Dordrecht.

Whitton, B. A. & Potts, M. (2012) Introduction to the cyanobacteria. In Whitton, B. A. (ed.), *Ecology of Cyanobacteria II.* Springer, Dordrecht, pp. 1–13.

Wilkinson, D. M. (1994) A review of the biogeography of the protozoan genus *Nebela* in the southern temperate and Antarctic zones. *Area* 26, 150–157.

Wilkinson, D. M. (2000) Running with the Red Queen: reflections on 'Sex versus non-sex versus parasite'. *Oikos* 91, 589–596.

Wilkinson, D. M. (2001) What is the upper size limit for cosmopolitan distribution in free-living microorganisms? *Journal of Biogeography* 28, 285–291.

Wilkinson, D. M. (2005) Brown knew the particles were smaller than pollen. *Nature* 434, 137.

Wilkinson, D. M. (2007) Do we need to worry about the conservation of micro-organisms? In Midgley, M. (ed.), *Earthy Realism: the Meaning of Gaia.* Imprint Academic, Exeter, pp. 52–59.

Wilkinson, D. M. (2010) Have we underestimated the importance of humans in the biogeography of free-living terrestrial microorganisms? *Journal of Biogeography* 37, 393–397.

Wilkinson, D. M. (2021) *Ecology and Natural History.* New Naturalist 143. Collins, London.

Wilkinson, D. M. (2022) Testate amoebae: a beginner's guide to a fascinating microworld. *British Wildlife* 34, 96–104.

Wilkinson, D. M. (2023) *The Fundamental Processes in Ecology: Life and the Earth System,* 2nd edition. Oxford University Press, Oxford.

Wilkinson, D. M. (2025) Ninety years of ecosystems. *British Wildlife* 37, 165–170.

Wilkinson, D. M. & Mitchell, E. A. D. (2010) Testate amoebae and nutrient cycling with particular reference to soils. *Geomicrobiology Journal* 27, 520–533.

Wilkinson, D. M. & Smith, H. G. (2006) An initial account of the terrestrial protozoa of Ascension Island. *Acta Protozoologica* 45, 407–413.

Wilkinson, D. M., Koumoutsaris, S., Mitchell, E. A. D. & Bey, I. (2012) Modelling the effect of size on aerial dispersal of microorganisms. *Journal of Biogeography* 39, 89–97.

Wilkinson, D. M., Creevy, A. L. & Fisher, J. (2017) Into the great unknown: the microbial diversity of a nature reserve. *British Wildlife* 29, 3–8.

Wilkinson, D. M., Mariani, M. & Cadd, H. (2025) Testate amoebae in bryophyte-rich biocrusts from Murray-Sunset National Park, Victoria. *The Victorian Naturalist* 142, 36–41.

Williams, P. J. le B. & Ducklow, H. W. (2019) The microbial loop concept: a history 1930–1974. *Journal of Marine Research* 77, 23–81.

Willmot, A. & Moyes, N. (2015) *The Flora of Derbyshire.* Pisces Publications, Newbury.

Wilson, C. G. & Sherman, P. W. (2010) Anciently asexual bdelloid rotifers escape lethal fungal parasites by drying up and blowing away. *Science* 327, 574–576.

Wilson, E. O. (1994) *Naturalist.* Island Press, Washington, DC.

Wolpert, L. & Szathmáry, E. (2002) Evolution and the egg. *Nature* 420, 745.

Wooton, D. (2015) *The Invention of Science.* Allen Lane, London.

Yang, J., Smith, H. G., Sherratt, T. N. & Wilkinson, D. M. (2010) Is there a size limit for cosmopolitan distribution in free-living microorganisms? A biogeographical analysis of testate amoebae from polar areas. *Microbial Ecology* 59, 635–645.

Yeates, G. W. & Foissner, W. (1995) Testate amoebae as predators of nematodes. *Biology and Fertility of Soils* 20, 1–7.

Zachar, I., Szilágyi, A., Számado, S. & Szathmáry, E. (2018) Farming the mitochondrial ancestors as a model of endosymbiotic establishment by natural selection. *Proceedings of the National Academy of Sciences of the USA* 115, E1504–E1510.

Zhang, Z. and 9 others (2024) New species and records expand the checklist of cellular slime molds (Dictyostelids) in Jilin Province, China. *Journal of Fungi* 10, 834.

Index

Note: Page numbers in **bold** refer to information contained in figure captions and figures.
Page numbers in *italics* refer to information contained in boxes or tables.

SPECIES INDEX

GENERAL INDEX

The New Naturalist Library

1. *Butterflies* — E. B. Ford
2. *British Game* — B. Vesey-Fitzgerald
3. *London's Natural History* — R. S. R. Fitter
4. *Britain's Structure and Scenery* — L. Dudley Stamp
5. *Wild Flowers* — J. Gilmour & M. Walters
6. *The Highlands & Islands* — F. Fraser Darling & J. M. Boyd
7. *Mushrooms & Toadstools* — J. Ramsbottom
8. *Insect Natural History* — A. D. Imms
9. *A Country Parish* — A. W. Boyd
10. *British Plant Life* — W. B. Turrill
11. *Mountains & Moorlands* — W. H. Pearsall
12. *The Sea Shore* — C. M. Yonge
13. *Snowdonia* — F. J. North, B. Campbell & R. Scott
14. *The Art of Botanical Illustration* — W. Blunt
15. *Life in Lakes & Rivers* — T. T. Macan & E. B. Worthington
16. *Wild Flowers of Chalk & Limestone* — J. E. Lousley
17. *Birds & Men* — E. M. Nicholson
18. *A Natural History of Man in Britain* — H. J. Fleure & M. Davies
19. *Wild Orchids of Britain* — V. S. Summerhayes
20. *The British Amphibians & Reptiles* — M. Smith
21. *British Mammals* — L. Harrison Matthews
22. *Climate and the British Scene* — G. Manley
23. *An Angler's Entomology* — J. R. Harris
24. *Flowers of the Coast* — I. Hepburn
25. *The Sea Coast* — J. A. Steers
26. *The Weald* — S. W. Wooldridge & F. Goldring
27. *Dartmoor* — L. A. Harvey & D. St Leger Gordon
28. *Sea Birds* — J. Fisher & R. M. Lockley
29. *The World of the Honeybee* — C. G. Butler
30. *Moths* — E. B. Ford
31. *Man and the Land* — L. Dudley Stamp
32. *Trees, Woods and Man* — H. L. Edlin
33. *Mountain Flowers* — J. Raven & M. Walters
34. *The Open Sea: I. The World of Plankton* — A. Hardy
35. *The World of the Soil* — E. J. Russell
36. *Insect Migration* — C. B. Williams
37. *The Open Sea: II. Fish & Fisheries* — A. Hardy
38. *The World of Spiders* — W. S. Bristowe
39. *The Folklore of Birds* — E. A. Armstrong
40. *Bumblebees* — J. B. Free & C. G. Butler
41. *Dragonflies* — P. S. Corbet, C. Longfield & N. W. Moore
42. *Fossils* — H. H. Swinnerton
43. *Weeds & Aliens* — E. Salisbury
44. *The Peak District* — K. C. Edwards
45. *The Common Lands of England & Wales* — L. Dudley Stamp & W. G. Hoskins
46. *The Broads* — E. A. Ellis
47. *The Snowdonia National Park* — W. M. Condry
48. *Grass and Grasslands* — I. Moore
49. *Nature Conservation in Britain* — L. Dudley Stamp
50. *Pesticides and Pollution* — K. Mellanby
51. *Man & Birds* — R. K. Murton
52. *Woodland Birds* — E. Simms
53. *The Lake District* — W. H. Pearsall & W. Pennington
54. *The Pollination of Flowers* — M. Proctor & P. Yeo
55. *Finches* — I. Newton
56. *Pedigree: Words from Nature* — S. Potter & L. Sargent
57. *British Seals* — H. R. Hewer
58. *Hedges* — E. Pollard, M. D. Hooper & N. W. Moore
59. *Ants* — M. V. Brian
60. *British Birds of Prey* — L. Brown
61. *Inheritance and Natural History* — R. J. Berry
62. *British Tits* — C. Perrins
63. *British Thrushes* — E. Simms
64. *The Natural History of Shetland* — R. J. Berry & J. L. Johnston
65. *Waders* — W. G. Hale
66. *The Natural History of Wales* — W. M. Condry
67. *Farming and Wildlife* — K. Mellanby
68. *Mammals in the British Isles* — L. Harrison Matthews
69. *Reptiles and Amphibians in Britain* — D. Frazer
70. *The Natural History of Orkney* — R. J. Berry

71. *British Warblers* — E. Simms
72. *Heathlands* — N. R. Webb
73. *The New Forest* — C. R. Tubbs
74. *Ferns* — C. N. Page
75. *Freshwater Fish* — P. S. Maitland & R. N. Campbell
76. *The Hebrides* — J. M. Boyd & I. L. Boyd
77. *The Soil* — B. Davis, N. Walker, D. Ball & A. Fitter
78. *British Larks, Pipits & Wagtails* — E. Simms
79. *Caves & Cave Life* — P. Chapman
80. *Wild & Garden Plants* — M. Walters
81. *Ladybirds* — M. E. N. Majerus
82. *The New Naturalists* — P. Marren
83. *The Natural History of Pollination* — M. Proctor, P. Yeo & A. Lack
84. *Ireland: A Natural History* — D. Cabot
85. *Plant Disease* — D. Ingram & N. Robertson
86. *Lichens* — Oliver Gilbert
87. *Amphibians and Reptiles* — T. Beebee & R. Griffiths
88. *Loch Lomondside* — J. Mitchell
89. *The Broads* — B. Moss
90. *Moths* — M. Majerus
91. *Nature Conservation* — P. Marren
92. *Lakeland* — D. Ratcliffe
93. *British Bats* — John Altringham
94. *Seashore* — Peter Hayward
95. *Northumberland* — Angus Lunn
96. *Fungi* — Brian Spooner & Peter Roberts
97. *Mosses & Liverworts* — Nick Hodgetts & Ron Porley
98. *Bumblebees* — Ted Benton
99. *Gower* — Jonathan Mullard
100. *Woodlands* — Oliver Rackham
101. *Galloway and the Borders* — Derek Ratcliffe
102. *Garden Natural History* — Stefan Buczacki
103. *The Isles of Scilly* — Rosemary Parslow
104. *A History of Ornithology* — Peter Bircham
105. *Wye Valley* — George Peterken
106. *Dragonflies* — Philip Corbet & Stephen Brooks
107. *Grouse* — Adam Watson & Robert Moss
108. *Southern England* — Peter Friend
109. *Islands* — R. J. Berry
110. *Wildfowl* — David Cabot
111. *Dartmoor* — Ian Mercer
112. *Books and Naturalists* — David E. Allen
113. *Bird Migration* — Ian Newton
114. *Badger* — Timothy J. Roper
115. *Climate and Weather* — John Kington
116. *Plant Pests* — David V. Alford
117. *Plant Galls* — Margaret Redfern
118. *Marches* — Andrew Allott
119. *Scotland* — Peter Friend
120. *Grasshoppers & Crickets* — Ted Benton
121. *Partridges* — G. R. (Dick) Potts
122. *Vegetation of Britain & Ireland* — Michael Proctor
123. *Terns* — David Cabot & Ian Nisbet
124. *Bird Populations* — Ian Newton
125. *Owls* — Mike Toms
126. *Brecon Beacons* — Jonathan Mullard
127. *Nature in Towns and Cities* — David Goode
128. *Lakes, Loughs and Lochs* — Brian Moss
129. *Alien Plants* — Clive A. Stace and Michael J. Crawley
130. *Yorkshire Dales* — John Lee
131. *Shallow Seas* — Peter J. Hayward
132. *Falcons* — Richard Sale
133. *Slugs and Snails* — Robert Cameron
134. *Early Humans* — Nicholas Ashton
135. *Farming and Birds* — Ian Newton
136. *Beetles* — Richard Jones
137. *Hedgehog* — Pat Morris
138. *The Burren* — David Cabot & Roger Goodwillie
139. *Gulls* — John C. Coulson
140. *Garden Birds* — Mike Toms
141. *Pembrokeshire* — Jonathan Mullard
142. *Uplands and Birds* — Ian Newton
143. *Ecology and Natural History* — David M. Wilkinson
144. *Peak District* — Penny Anderson
145. *Trees* — Peter A. Thomas
146. *Solitary Bees* — Ted Benton & Nick Owens
147. *Shieldbugs* — Richard Jones
148. *Ponds, Pools and Puddles* — Jeremy Biggs & Penny Williams
149. *Stoats, Weasels, Martens and Polecats* — Jenny MacPherson
150. *Exmoor* — Flemming Ulf-Hansen

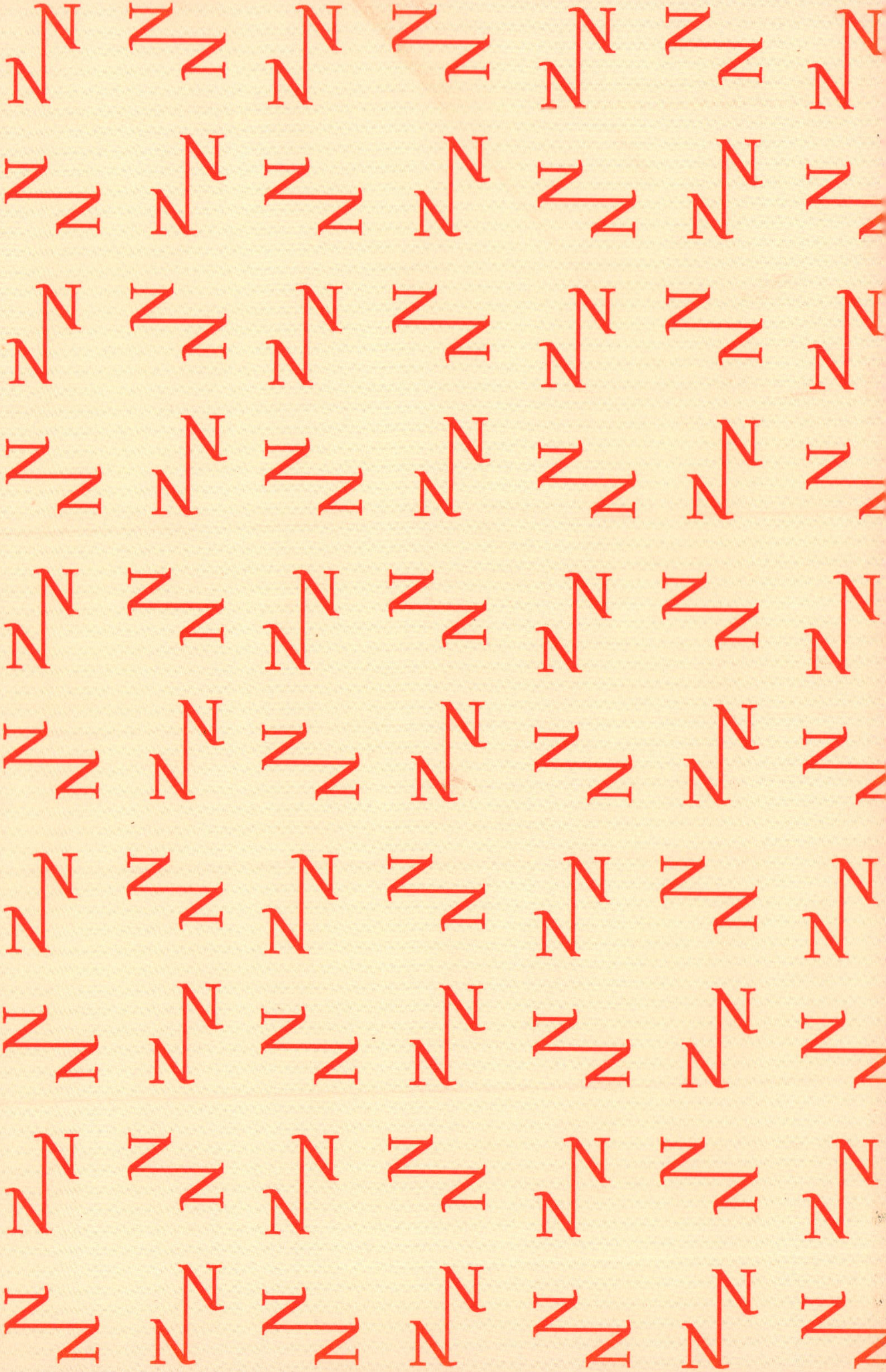